Joseph Franklin Perry

Dogs: Their Management and Treatment in Disease

Vol. 2

Joseph Franklin Perry

Dogs: Their Management and Treatment in Disease
Vol. 2

ISBN/EAN: 9783337778927

Printed in Europe, USA, Canada, Australia, Japan

Cover: Foto ©berggeist007 / pixelio.de

More available books at **www.hansebooks.com**

DOGS:

THEIR

MANAGEMENT AND TREATMENT

IN

DISEASE.

A STUDY OF THE THEORY AND PRACTICE OF CANINE MEDICINE.

———

By ASHMONT.

———

PUBLISHED BY THE AUTHOR.
1885.

PREFACE.

The rapid advancement in science has wrought many changes in the principles and practice of medicine; added new remedies to the resources of the physician; exploded many old theories, and dissipated countless fancies and prejudices. Well worn paths have been left, new fields traversed, and discoveries made, which aid not only in the cure, but in the prevention of disease. Important lessons have been taught, and the limits of knowledge extended, by trials of endurance which have developed unsuspected latent powers, both in man and animals. Further progress will yet be made, of which speculation is fruitless.

To place before the reader a faithful exposition of medicine, as it exists at the present time, is the purpose of the author. No finished treatise is attempted, replete with scientific technicalities, calculated to confuse and mystify, but in simple, comprehensive language and expression, on these pages are described the prominent appearances in disease, and the manner of applying remedies, sustained largely by the results of personal observation.

This book is submitted with the ardent hope it may be found a source of instruction, and aid in the care of man's true and faithful friend, the Dog.

ASHMONT.

CONTENTS.

DOGS:

THEIR

MANAGEMENT AND TREATMENT

IN DISEASE.

CHAPTER I.

INTRODUCTION.

As the object of this work is to treat of the dog in a state of disease, the author has deemed it expedient to confine himself to those essential considerations in keeping with his purpose, reserving for a future effort the subject of Breeding and attendant interests.

The introduction of *general management* which might seem extraneous, is yet pursuant of this design, on the assumption that a knowledge of the *prevention* of disease, is intimately asssociated with the study of *pathology*, and holds a place scarcely inferior to that of a familiarity with the principles and practice of medicine.

Many diseases are preventable by removing or obviating their causes; reforms in ventilation, cleanliness and diet, there is reason to believe, accomplish much, and are among the most important modes by which diseases may be prevented.

The care of puppies from the period of weaning, is treated of in a general way. In the management of dogs after maturity is reached, minor details are dwelt upon, it being assumed that, to some at least, their consideration may prove of value.

The number of drugs capable of producing death, under certain conditions, is almost limitless, as medicines, possessing sufficient activity to be of much value, are always poisons in inordinate or excessive quantities.

In the chapter devoted to the subject, the action of dangerous agents, more generally encountered, are discussed.

In the consideration of the different diseases, *causes, symptoms, treatment and prognosis* are each of recognized importance, and *diagnosis* if necessary to discriminate from one of several diseases with which it has more or less symptomatic phenomena in common.

To dwell on *pathological anatomy*, or the appearances in a morbid or diseased state, must necessarily confuse the reader and reference excepting in rare instances is avoided.

In the arrangement of diseases convenience is studied. Under *general diseases* are considered all unclassified.

It is implied that *operative surgery* is beyond the capacity of many, and as an exhaustive treatise on the subject would prove of but little value, only the simpler operations, and treatment in cases of *emergency* are considered.

THE CARE OF PUPPIES AFTER WEANING.

One of the greatest essentials in the rearing of puppies, especially the case in large breeds, is sufficient exercise. Under no consideration, chain them, but allow almost absolute freedom of movement. If kept in a pen, have it large and so constructed that they can lie down or stand on all fours and look between the boards; for standing on their hind legs to look over, would invite weak backs and cow-hocks. The sleeping quarters should be kept scrupulously clean, frequently whitewashed, and free from draughts. In the warmer weather, pine shavings or sawdust is the best bedding; in the winter, clean straw, changed frequently.

In feeding puppies, their discharges should be watched, and the diet varied as the need manifests itself. Milk is one of the most important articles of food in the earlier weeks after weaning. It may be frequently given, some caution being used, however. It should be old or skimmed, never fed fresh from the cow. Three or four times a week, sour milk may be given, as it is an enemy to worms. It will be well to add limewater to sweet milk when used, in the proportion of nearly one-sixth part.

Sometimes when feeding milk freely to very young puppies, diarrhœa manifests itself. A change must at once be made to beef tea and broths, into which may be broken crackers or bread. Too great importance cannot be given to the early administration of meat, which must be fresh and lean. Unsound food should never be used in any form. Meat may be given raw or cooked, small quantities of the former at first, and when the age of three or four months is reached, let it be scraped or finely cut, and given after other food has been eaten.

Meat may be cooked alone or with vegetables, onions in small quantities being especially healthy and palatable, then removed, and in the liquor, oatmeal, indian meal, or wheat middlings, be well boiled and thoroughly cooked. Then the meat can be chopped up and added with the crushed vegetables. Rice and pearl-barley may be given occasionally.

Liver, from its laxative action, must be given guardedly, if at all. Large bones may frequently be supplied, but beware of small bones that may possibly sliver, as many a valuable dog has been killed by them. Remem-

ber always the great secrets in feeding are — wholesome food and a variety
in diet. Where only a few dogs are kept, the "scraps" from the table,
from their varied nature, cannot be improved upon. Mix potatoes and
other vegetables with a little gravy and small pieces of meat. Until six
months old, feed three times a day, then twice daily until maturity is
reached.

Remain with the pup while eating, and as soon as he turns from his food
remove it at once. Never urge a dog to eat, if he shows from his manner
a loss of appetite. Prepare something else and tempt him with it and if he
still declines, wait until his next regular time of feeding and try again.

As a rule, give as little medicine as possible to puppies. Probably the
more common affection observed among them is indigestion in some form
or other. As castor oil or syrup of buckthorn can do no harm, if an odor to
the breath, a hot nose, sluggish manner, or disinclination to eat is noted,
give either, and watch the discharges. If very offensive, or undigested
matter appears in them, the cause is apparent. Stop all food for a time,
and when next given, select only the simplest and most easily digested,
milk and lime-water preferred.

Vegetables have a tendency to purify the blood and keep the bowels in
good condition. When these cannot be given, other means should be
employed in the form of a mild physic. It is a good plan to give a dose of
sulphur, either alone or with powdered magnesia, if a laxative is needed,
mixed with the food once or twice a week in summer, and less often in
colder weather. Ordinary dose, one-half a teaspoonful of each.

Diarrhœa is usually the result of indigestion. Often nature's effort to
throw off the disease, in many cases regulating the diet, will suffice; giving
porridge made of boiled milk, and flour baked until slightly brown. If the
attack is severe, first imitate nature by giving a dose of castor oil, it being
desirable that the contents of the intestines be effectually removed, in order
to prevent their continued passage over the inflamed surface, and to
secure, as far as may be, that important end in the treatment of all inflammations, viz., quietude of the inflamed parts. Paregoric, as the simplest
form of opiate, may then be given in one or two teaspoonful doses.

Worms are a common enemy to pups, and the giving of sour milk will
do much to prevent their formation. Their presence may be suspected
when bloating beyond the fulness that the food taken would be likely to
produce. For worms in puppies three months old, give one or two grains
of santonine in butter after having fasted twelve hours; follow this dose
three hours later with one of castor oil.

Fleas are a constant source of annoyance. Washing young puppies is
never admissible. Combing with a fine comb dipped in kerosene oil, or
rubbing flea powder into the hair, are the means to be employed.

GENERAL MANAGEMENT.

FEEDING.

Pages have been written on the origin and ancient history of the dog, theories advanced, speculations urged, and yet we still remain in complete ignorance on the subject. Some contend, in the anatomical structure of certain parts that he resembles the wolf. There are however, differences to reconcile, before this supposition can be accepted. Others have traced to the fox, and still others to the jackal, each party being satisfied in their own minds of the correctness of their theories, and yet the same differences remain.

A knowledge of the primitive state and habits of the animal might aid us much on this subject, but that being denied us, we are forced to turn to the anatomical mechanism for guidance.

A study of the digestive organs of the dog proves him a carnivora, or flesh eater, and meat is unquestionably his natural food. The traditional prejudice, that its free use perverts the scent of sporting dogs, generates mange, and injures the dispositions of the animals, making them ugly and savage, is rapidly becoming "among the things that were."

The instinctive desire of animals for certain substances is the surest indication that they are in reality required for the nutritive process; on the other hand, the indifference or repugnance manifested for injurious and valueless substances is an equal evidence of their unfitness as articles of food.

Starch is very abundant in corn, wheat, rye, oats, and rice, and in almost all vegetable substances used as food, and during the process of digestion is converted into sugar. That these articles alone are an unfit diet, is proven by the experiments of Magendie, who found that dogs fed exclusively on starch or sugar, perished after a short time with symptoms of profound disturbance of the nutritive functions. On the other hand experiments have been made by Claude Bernard and others, proving that carnivorous animals can be kept for an indefinite time, exclusively on a diet of flesh, and yet the body be maintained in a perfectly healthy condition.

Dogs have no desire for vegetables, no natural cravings for them. To be so fed, or on puddings, bread and starchy food continuously, only invites debilitated constitutions and attendant diseases.

The proper food for a dog is Meat, varied in any way preferred, but it should be the principal article of diet.

The question arises whether meat should be fed cooked or raw. Both have advantages. Certain constituents are lost in the process of cooking. The muscular fibres are almost always more or less hardened by boiling or roasting, but at the same time, the fibrous tissue by which they are held together is gelatinized and softened, so that the muscular fibres are more easily separated from each other, and more readily attacked by the digestive fluids. The organic substances contained in meat, which are insipid in the raw state, acquire by the action of heat and cooking, a peculiar and agreeable flavor. This flavor excites the appetite and stimulates the flow of the digestive fluids, and renders in this way the process of digestion more easy and expeditious.

In cooking meat, it is desirable that it be rendered palatable, and the flavor varied from time to time. This can be done by the addition of vegetables, which may be well crushed after a thorough boiling.

As the feeding of meat alone would prove an expensive diet in large kennels, different kinds of meal may be added to thicken the broths, but the quantity of starchy food must be very small. The writer is firm in his conviction, that its continued use does as much to propagate mange as any other abuse can.

Sheep's heads and the feet of animals may be boiled to a jelly. They are highly nutritious, and dogs are very fond of them.

Beef flour is a valuable substitute for flesh, is excellent for puppies and matured dogs alike; it can be added to boiling water, a strong broth made, and then thickened with a little bread, crackers, or meal.

Pork scraps, in pressed cakes, keep for an indefinite time, and may be fed once or twice a week; they should be soaked for some hours until soft, and then boiled.

Milk, sweet or sour, can be given freely, and is valuable for a change. Dog biscuits, so called, have within a few years become very popular in large kennels; when honestly prepared, they are an excellent diet. The manner of feeding them should be frequently changed; at times given dry, at others soaked in milk or broths, or crushed with vegetables.

Bones, the dog's tooth brush, are an essential. By their use the muscles of the jaws are kept strong, and the pleasure they give the animals is very evident; their value may be lessened however, by too frequent feeding, as many a broken well worn tooth testifies.

While the use of vegetables adds flavor to the food and excites the appetite of the animal, it has still another important action as a gentle laxative. Liver has the same property, and the conditions of the bowels should direct its use.

Among the medicinal agents often administered with the food, sulphur may be properly referred to while on the subject of feeding. It is laxative and resolvent, and is supposed to be rendered soluble by the soda of the bile. It is thrown out of the system by the bowels and through the skin, and for that reason has an influence on cutaneous diseases. Its occasional use can do no harm, and unless a perfect condition of the animal contra-indicates, it may be given once or twice a week in summer, and once or twice a month in colder weather.

The dose for a fifty pound dog is one teaspoonful, and may be mingled with the food or given in milk.

Many authors have dwelt upon the dirt-eating propensity of dogs, and concluded therefrom, that the proper method of feeding is to throw the food on to the ground. The writer is of the opinion that if such a propensity exists, it is wiser to leave the dogs to nature's promptings and let them eat dirt when they choose, rather than insist upon its being eaten at every meal. He therefore prefers to feed his animals from clean dishes, and the cleaner the better.

The greatest importance is attached to the quality of the food given. It should be of the very best, and no tainted meat, mouldy bread, or food changed in its integrity, should ever be used in any form. Many people consider decomposing meat better for a dog, and attempt to fortify their arguments, by illustrating with the fact, that the animal will occasionally bury a bone or part of his food, to be dug up and eaten later. The fact is true, but the inference an ancient absurdity. A dog never buries meat when he is hungry, but after a hearty meal will often do so, proving our supposition that it is his natural food, an evidence of his great craving, which he will eat in any form, no matter how badly decayed. Give him sufficient fresh meat and he will never seek carrion.

Chloride of Sodium or common salt, some authors state, "is poison to a dog." In this the writer cannot agree, but on the contrary deems it essential to perfect health. It is found throughout the tissues and fluids of the body, exerts an important influence on the solution of various other ingre-dients with which it is associated, and the blood globules are maintained by it in a state of integrity. Since common salt is so universally present in all parts of the body, it is an important ingredient of food. It occurs in all animal and vegetable food as well, though in smaller amount in the latter, and acts in a favorable manner as a condiment, by exciting digestion, and assisting in this way the solution of the food.

In connection with the subject of feeding, reference to the appetite is pertinent. Remember dogs will eat when hungry. For them to refuse a meal occasionally, means nothing. Evidence is not wanting where they have passed days and even weeks in confinement without food, and yet recovered without injury from their long fast. If food is placed before the animal and he turns from it, something else should be prepared to tempt him. If he still declines, it should be removed and another attempt be made at his next regular time of feeding. If a dog is in good health, simply dainty, this method repeated again and again, if needed, will prove effectual, and he will finally eat whatever may be offered him.

Regularity in the hour of feeding should be observed. The number of times to feed is a disputed question, with the weight of authority on the side of once a day, at night, and all they can then eat. Owners must decide for themselves, influenced by what has been their custom, the condition of their animals, and the amount of work required of them. Where only one dog is kept, no doubt in many instances it is over fed, especially if allowed the freedom of the kitchen While the fact is apparent, that once a day is all a dog actually needs to be fed, the question arises: would it not be wiser to give him a little less at night, and a mere trifle in the morning? It would seem thereby that the dangers of indigestion would be lessened.

It might be said that even of greater importance than good wholesome food, is a constant supply of good fresh water. The system suffers more rapidly when entirely deprived of fluids than when the solid food only is withdrawn. Magendie found in his experiments on dogs subjected to fasting, that if the animals were supplied with water alone, they lived six, eight, and even ten days longer than if they were deprived the same time of both solid and liquid food.

To recapitulate: the essentials in feeding are meat, fresh and wholesome, varied from the raw by cooking, and the addition of other articles of food, to add to the flavor and tempt the appetite, together with a constant supply of fresh, pure, and clean water.

EXERCISE.

Exercise develops the muscles of the body, increases their functional activity, hastens the digestion of food, and renders its assimilation easier. It improves nutrition by elevating the temperature, opens the pores of the skin, stimulating its functions, and through the lungs, by the stronger movements excited, it favors the exhalation of impurities and excesses.

Too much importance cannot be attached to this subject. Keeping a dog confined with insufficient exercise is cruelty itself, which no excuse can justify. *Never chain a dog* should be the rule, but unfortunately it cannot always be observed. Certain methods may, however, be resorted to, and a confined animal have greater freedom and still be restrained.

A post can be set up as far from the kennel as the limits of the owner's premises will allow, a wire fastened to it and passed through a ring, the other end attached to the kennel and drawn taut. The dog can then be chained to the ring, which will slide freely the entire length of the wire, permitting greater latitude and the animal still be held in check. This means may be employed in cities, and where it is impossible to allow them to run at large.

When dogs are kept in yards, the plan adopted by the writer can be resorted to. Fences or hurdles can be built entirely across the inclosures. These must be placed low at first, simply high enough to force the animal to make an easy jump, then in three or four weeks raised a few inches, another interval allowed to pass, and raised again. The results from the employment of this means are admirable, especially in the case of dogs inclined to be weak in the loins, and needing development behind.

In conditioning dogs, especially for exhibitions, no better exercise can be recommended than slow walking; the distance being from two to four miles daily. By this means they will eat more, digest it better, and accumulate flesh more rapidly.

While exercise is imperative, extremes must be avoided and caution used. If a dog is habituated to following a team, there is less danger of over exertion; if not so accustomed, care must be observed lest the system be disordered by too great effort. In winter, if a hard run is taken and he returns in a heated condition, he must not be shut into a cold kennel, but be allowed to *cool gradually* in a warm room. Violent exercise must never be allowed after a hearty meal.

It is a wise plan to accustom dogs to climatic changes, yet they must be protected during the hotter days of summer, from the sun's intense heat, or the exposure will cause debility and often convulsions.

WASHING.

Washing is admissible in summer, and undoubtedly has its benefits in colder weather when all influences are favorable, and proper precautions observed; unfortunately, neglect results too seriously, and it cannot be insisted upon, that bathing is an absolute necessity for dogs.

Some breeds take to the water naturally and find pleasure in it, but the reluctance of others, and the disposition shown by them to conceal themselves when the preparation to wash commences, is an indication that the process will prove a hardship. Where it is possible, it is better to allow them access to a tub or cistern of water, and leave them to follow their natural promptings.

Very dirty coats, vermin, and cutaneous diseases sometimes make washing imperative. If these needs are not evident, it is wiser to do little or none of it, for care in the kennel arrangements and thorough grooming will obviate the necessity.

Washing in cold weather is extremely hazardous, and the animal is almost certain to take cold unless the greatest care is observed after the operation. He must be confined to the house until thoroughly dry, then judiciously exercised before returned to his kennel.

In warm weather the dangers are much less, yet care should be observed. The method of washing naturally suggests itself, and need not be dwelt upon. The purpose for which the animal is washed will direct the nature of the soaps to be used. If mere cleanliness of the coat is the object, champooing is the better means to employ.

Eggs broken and rubbed into the hair, then sponged off, leave it clean, soft, and glossy. The *sea-foam* champoo of barbers is also excellent as a stimulant and for cleanliness.

Fleas demand the use of soaps which destroy them. Cutaneous diseases call for *carbolic, sulphur,* and the varieties of soaps mentioned elsewhere under *skin diseases.*

If used properly there is little or no danger of absorption from the strong carbolic preparations; the operation must be rapidly performed, and by a thorough rinsing every trace of the soaps removed.

After washing, the animal should be thoroughly sponged, rubbed dry, and then properly cared for.

KENNEL ARRANGEMENTS.

In kennelling, the primary essentials are, good ventilation, perfect clean-liness, dryness, and freedom from draughts.

Less sickness undoubtedly occurs when dogs are inured to climatic changes. Those animals accustomed to out-door kennels, are usually far healthier than those that are allowed the freedom of the house. Again when a disease like distemper attacks them, it runs a shorter and much less severe course; their natures, more elastic, seem to throw off the in-firmities quicker. .

In winter dogs suffer as much, if not more, from artificial heat than they do from exposure to intense cold. Many are allowed in houses and around the stoves during the day, and are kennelled out at night, or shut into a cold stable, thus experiencing the greatest extreme in temperature; a fruitful source of disease.

To kennel out and constantly keep in the open air, the coarser long-haired breeds, seem the wiser. Their houses should be carefully con-structed, raised several inches from the ground, and the space beneath carefully boarded. These should, in winter, face the rising sun, and be provided with a glass front. The door may be hung on hinges, so that it can be opened by the dog from within or without, or it may be dispensed with, in which case a projection should be built around the door-way on the outside, to prevent rain from entering.

A room within a room in winter is the better. The inner or sleeping quarters can be easily constructed by partitioning off a part, leaving an opening as a passage for the animal, then laying a supplementary raised floor on which the bedding can be placed. This inner kennel can be ren-dered warm, snug, and free from all possible draughts, by covering the floor and sides with tarred paper.

Where dogs are housed at night in stables, barns, or large rooms, these same sleeping boxes should always be used in the colder weather.

The floors of the kennel should be covered with sawdust as it is the cheapest and cleanest. For bedding, straw in winter, and fine shavings in summer are preferred.

Cleanliness is of paramount importance. In this connection, the meth-ods employed by the writer may be of interest. Every week in warm and cold weather alike, in his kennels, the sawdust and bedding are removed, the doors and windows closed, and the building purified by introducing a kerosene stove, and burning sulphur thereon. The walls, even the floors if necessary, are then thoroughly washed with lime, and if the weather is damp or very cold, the stove is again introduced and the white-wash dried. After all dampness is removed, the floors are covered with a disinfectant, lime and carbolic acid more often used, tarred paper then laid down, on it sawdust, and then the bedding. It is important where several dogs are kept, that they be provided with separate kennels, at least no more than two be allowed to remain together, as infectious diseases are sometimes transmitted, in their earliest stages, and may become general before detection. This danger is especially great where the care of the animals is intrusted to servants.

GROOMING.

In the stable, one often hears the old adage, "a good grooming is better for a horse than a feeding." The same holds true in the kennel. The ancient Averrhœs gave the sum of the directions by Galen and other authorities, in his reference to friction applied to the human body. Strong friction, he said, braces and hardens the body; weak, rarefies and softens; moderate, in an intermediate degree. Besides, he added, hard friction diminishes obesity; moderate, on the other hand, tends to remove emaciation.

Not only is the health of a dog improved by grooming, but his changed appearance well repays the owner for the effort. Again in the operation, commencing cutaneous diseases and vermin are detected, suggesting the prompt application of remedies and much trouble and annoyance thereby averted.

In the use of a comb, care should be observed lest the skin be irritated and mange propagated.

A clean, fairly stiff hair brush, a bathing hair-glove, or a pad of woollen yarns, and a chamois skin are all the essentials. Their use can be followed by hand rubbing, an admirable method of adding a fine finish and producing a gloss.

VERMIN.

Fleas, more especially in warm weather, are the bane of a dog's existence, and only the most constant care and watchfulness can insure protection from this unmitigated nuisance.

There are literally two species of this parasite; the common-flea *(pulex irritans.)* and the sand-flea *(pulex penetrans.)* The former bites, producing papules or wheals; the latter bores into the skin, causing inflammation.

The *sand-fleas*, as the name implies, are common in sandy districts, and are very difficult to combat. Even when entirely removed, they are quite certain to return. Kennels infected, if situated in the sand, should be moved to a foundation of black earth. If that is impossible, earth should be drawn and the floors and ground surrounding be covered with it, then insect powder be blown freely into every crack and crevice, and rubbed well into the dogs; this will drive them away for a time, but the method must be repeated or they will return.

The *common-flea* needs much the same continued effort to suppress him. Grooming, combing, and washing will do much, but generally other means must be employed.

Whale oil is an effectual remedy but its use is an abomination; it must be thoroughly rubbed into the hair of the animal, from tip to tip, and allowed to remain on, some hours at least, during which time he must be kept in a warm room, as there is great danger of taking cold; then eggs may be used as a champoo and washed off, or soap and water alone, followed by a persistent combing, and after all a hard run be given.

Quassia chips boiled to a strong infusion, or in the form of tincture, when sponged on freely, will be found efficacious. *Carbolic acid* and soaps incorporated with the same are excellent, but must be used cautiously, or poison from absorption may result; a mixture of carbolic acid, two drachms of the crystals to a pint of water, is sufficiently strong. The *phenol sodique* is preferable to carbolic acid, as there is little danger of poisonous absorption, and may be used diluted with the same quantity of water.

The *sulphuret of lime* used with such success in Belgium for the itch, acts quite well as a cure of vermin; a weak solution of it can be made by boiling together the flowers of sulphur two pounds, unslaked lime one pound, water two gallons; slack the lime first then add the sulphur slowly, stirring well, and finally boil down to one gallon. Let the mixture cool and precipitate, then pour off the clear fluid and apply it freely to the hair of the dog, allowing it to dry on.

Balsam peru is an agent which has been, in times passed, much valued in parasitic troubles. Its efficacy, is unquestioned, but it is rather too expensive for general use.

Kerosene and *crude petroleum* are used advantageously in many kennels, the method employed being, to dip a comb into the oils and pass it persistently through the hair, being careful that little or none reaches the skin, which is likely to become irritated thereby.

Insect powder is an excellent remedy, and can be economically used, by laying the dog on a paper while being rubbed.

Lice and *wood-ticks* demand much the same treatment that fleas do, the preference being given to *petroleum oil*. A strong infusion of the seeds of the *stavesacre* is a sure remedy, but they are not easy to procure. Many authors advise the use of *mercury* in some form, the white precipitate the more common. It should not be employed until all the simpler remedies have failed, and then be used with great caution, the dog being securely muzzled.

Where vermin have become troublesome, the kennels should be thoroughly cleaned, bedding destroyed, and sulphur burned, after which, all parts should be faithfully whitewashed.

The continued scratching of animals causes eczema which must be treated, after the vermin is removed, with external applications of *cod-liver oil, sulphur* and *lard,* or *phenol sodique,* and possibly the employment of the usual *mange* remedies will be demanded.

CHAPTER II.

ANTIDOTES AND TREATMENT

IN

CASES OF POISONING.

It would seem, that in the treatment of dogs in disease, some people draw from a fund of accumulated and transmitted ignorance, and display the least possible common sense and judgment. Remedies of the greatest efficacy and virtue, become exceedingly dangerous in the hands of the careless and incompetent, and many dogs are sacrificed by man's stupidity, as well as destroyed by that fiend incarnate, the dog poisoner.

No where will that old saying better apply than in connection with the use of medicine, " a little knowledge makes men foolish."

An article lately appeared in a prominent sportsman's paper, under the heading, "Treatment of Poisoned Dogs," which well illustrates this. It said, " The lives of many valuable dogs can be saved by the prompt use of a very simple remedy. As soon as you know a dog has been poisoned, inject about an ounce of hydrate of chloral into his back with a hyperdermic syringe, the quantity to be governed by the size of the dog and severity of his symptoms. As long as there is life in him do not despair. I have known dogs to have been saved by this treatment when in the death throes."

Nothing could be either more absurd, or more dangerous than this advice. One ounce of chloral hydrate is *four hundred and eighty grains;* the dose of this agent is from *five* to *twenty grains.* The rule when medicine is administered subcutaneously is *one half* the quantity given by the mouth. In cases of strychnine poisoning, the need is more urgent, and the usual doses can be safely increased; *twenty grains* however would be quite enough to introduce hyperdermically, as it acts quickly, and the need of a larger dose would be readily apparent. The adviser says " the quantity to be governed by the size of the dog." It is presumed it would matter little, were anything near four hundred and eighty grains administered. Again to " inject into the back," an abscess would surely result, leaving an unsightly scar. While chloral hydrate is indicated in cases of poisoning by *strychnine* and other drugs producing convulsions, its use would *surely prove fatal* in poisoning by *narcotics.*

Not alone in the care of animals, is shown this same stupid reasoning. The writer recalls a case he once treated, that of a woman who had effectually used *creosote* on a pledge of cotton pressed into an aching tooth. Shortly after on being attacked with *earache*, reasoning that the remedy, if a success in one instance, must surely be good in another, had a quantity dropped into her ear; the result can be imagined.

The writer had a valuable horse overcome by the heat. His driver much alarmed, was ready to do anything and everything advised. A bystander recommended *an ounce of the tincture of aconite root*, which was at once secured and administered. At least *eight hundred drops* were given of the drug, whereas *ten to fifteen drops* would have been reasonably large. Man's stupidity will never allow the limit of illustrations of this character to be reached.

While on the subject of poisons, certain rules in the use of medicines are appropriate.

Never undertake to prepare complicated prescriptions, but depend upon a competent and reliable druggist.

Protect your labels. If one is lost sacrifice the contents of the bottle rather than be in doubt.

Use great care in dropping medicine. When uncertain about the accuracy of the dose, throw it away and drop again.

Remember that medicine *can be repeated* if necessary, but *cannot be recalled* after once given. Give *too small* rather than *too large* doses.

All drugs require a certain time to act in, and must not be repeated until a proper interval has been allowed.

There is no protection from the poisoning fiend. Apothecaries are by law forbidden to sell poisons. They can easily be obtained however without legal responsibility. Pills of strychnine, nux vomica, arsenic etc., each containing the proper dose can be bought without suspicion, implying they are for personal use, several of them pressed into meat, thrown into the vicinity of the hated dog, and the hellish purpose be accomplished.

One author has stated, "as a general rule for distinguishing between the evidence of poisoning and the symptoms of disease, the suddenness of the attack must weigh largely; and by tracing where the dog has been, and what he has or is likely to have picked up, a pretty accurate conclusion can be arrived st.

This is about all that can be said on this subject. At the same time, it is essential to remember that there are many exceptions to this rule.

Some agents are more readily absorbed than others. Poisons taken into the stomach when empty, necessarily act much more speedily than when full, thus if that organ is loaded the appearance of the symptoms may be delayed some hours. Sleep may retard the action of some agents. It must be also remembered that there are many diseases which commence suddenly, and rapidly run to a fatal termination. Internal hemorrhages, severe inflammations of the stomach or intestines, often set in suddenly and might be taken for poisoning.

The process of diagnosis by elimination, described elsewhere, will aid much in cases of suspected poisoning.

It would be advisable where valuable dogs are owned, that a few remedies be prepared, and kept in anticipation of possible poisoning. *Sulphate of zinc* for an emetic, *laudanum* or *paregoric* for pain, *chloral hydrate* for convulsions (mixture of one drachm of the drug to an ounce of water). These, with a glass syringe holding from one to two tablespoonfuls, are of great use in emergences where delay is fatal.

In cases of poisoning, the methods of treatment indicated are : — Get rid of the Poison — Stop its action — Remedy the mischief it has done.

A consideration of the more common poisons and their antidotes, is all that space permits. Aside from drugs, other means are sometimes employed by the destroyers. *Powdered glass* is often used, and a *sponge* compressed while wet, then tightly rolled with twine, dried, and introduced into a piece of meat, is another method, producing certain death by intestinal obstruction. When glass is known to have been taken into the stomach, never give cathartics, but exclude drinks and give solid food with the hope of enveloping the particles, and thereby protecting against internal laceration.

Arsenic:—Symptoms; constant hawking, caused by burning pain in the throat; great thirst, tongue and mucous membrane of the mouth becomes red and swollen; abdomen enlarged, hard, tense, and painful to the touch; severe griping pains; vomiting and purging of brown or bloody matter. Symptoms grow rapidly worse, skin at first hot, but later cold and clammy, prostration, paralysis, convulsions, and death.

Treatment ; — If vomiting is not free, give an emetic, then quickly pour into the animal what may be within reach, either milk, flour and water, magnesia and oil, or oil and lime water, and send at once for the only true antidote, the *hydrated peroxide of iron*, for which, if the druggist has not all prepared, he can make a substitute on the instant, by adding to diluted tincture of iron enough bicarbonate of soda, or aqua ammonia, to saturate it; give freely of this.

Strychnine and Nux Vomica. —Symptoms; at first restlessness, then pain, as evinced by the dog's sharp cries, followed by twitchings of the muscles, jerkings of the head, snapping of the jaws, and foaming at the mouth; then convulsions, which may intermit for a short time, during the interval the animal uttering sharp, shrill, ear piercing cries, to be stifled by a recurring spasm.

Treatment;— If possible, give an emetic. This can, however, rarely be done after the convulsive stage has set in. Then the dependence must be on *chloral hydrate,* from twenty to thirty grains, administered by the rectum. Of a mixture of this agent, (one drachm to an ounce of water), a tablespoonful contains thirty grains of the chloral. After giving the injection, pressure should be made against the anus, to prevent its being evacuated. If in twenty minutes the convulsions have not ceased, repeat the injection, dose being the same. When the convulsions are finally controlled, allow the animal to sleep as long and quietly as possible, and when he arouses up and shows a disposition to move about, give, every three or four hours, one teaspoonful of *aromatic spirit of ammonia* well diluted, and continue at intervals until perfectly conscious.

Carbolic Acid:— Symptoms; Great depression, trembling, shivering, and loss of motion, indicative of approaching paralysis, diarrhœa, the discharges at times bloody.

Treatment;— Hot mustard bath, friction, and stimulants of *brandy* and *ammonia.*

Phosphorus:— Symptoms; Burning pain in the throat and stomach, vomiting, purging, great inflammation and tenderness of the abdomen, convulsions.

Treatment;— Give promptly an emetic, followed with either *magnesia, chalk,* or *whiting* in water. *Avoid oils,* as they dissolve the phosphorus.

Mercury:— Symptoms; Acts much like *arsenic,* but quicker and more violent, corrosion of mouth, burning of throat, distress in stomach and bowels, excessive thirst, vomiting and purging of bloody mucus, skin cold and clammy, convulsions.

Treatment;— If vomiting does not occur, induce it with an emetic; give freely white of eggs with milk, or flour and water, or flour and soft soap thinned with water. The *protosulphuret of iron* is an antidote, but is useless ten minutes after the poison has entered the stomach. Mercury, it must be remembered, occurs in many forms, the more common being *corrosive sublimate, red precipitate, white precipitate, cinnabar, vermillion, and cyanide of mercury.*

Lead Salts:— Symptoms; Depend on the mode of poisoning. In large doses, the usual symptoms are irritation, distress, and vomiting, colic, constipation, cramps, and paralysis.

Treatment;— Give *epsom salts* with milk and eggs freely.

GENERAL ANTIDOTE—In cases of poisoning, when the nature of the poison is unknown;—*calcined magnesia, powdered charcoal, sesqui oxide of iron,* equal parts of each in a sufficient quantity of water.

CHAPTER III.

—◆◇◆—

THE PRINCIPLES OF MEDICINE.

————◆◆————

PATHOLOGY.

If the term health expressed a well defined state, an absolute standard might be fixed. To define disease is equally as difficult. It is an absence, or deficiency of health, simply a transfer of the difficulty of definition.

If all the tissues and organs of the body are normal, if all the fluids are in no respect abnormal, if all the functions of the organical structure are completely and harmoniously performed, health undoubtedly exists. But this perfection of health is visionary and never actually exists. An examination of the healthiest would disclose some deviation, some change, and these deviations from the normal are not inconsistent with the evidences of health. Functions of different parts may be disordered to a certain extent, without sufficient disturbance to constitute disease.

No practical embarrassment can result from this difficulty, to draw the line with precision. Each owner will fix an individual standard for his animal. Variations from it will constitute disease.

In a general way, in a healthy dog, we note as follows:—Eyes bright, the white usually clear, the fine red lines seen at times having no significance, the lining of the lids a pink rose tinge. Nose, cold, moist and slippery, except when the animal sleeps, then often hot and dry. Coat soft, smooth, and in long haired dogs, glossy. Skin soft, easily moulded, and of a gentle heat. Tongue moist, pink in color, free from coating. Pulse, full and strong, ranging from eighty to one hundred, varying in different breeds and natures. The larger animals have a lower rate than the smaller, the nervous a higher than the less easily excited. Bowels; excretions vary with the food given, in consistency and color. They should be neither hard nor thin, free from undigested matter, and not markedly offensive in odor. Kidneys; urine, pale yellowish, abundant, freely and easily expelled.

DIAGNOSIS.

The term diagnosis, signifies the art of discriminating diseases, to determine their character and situation. It is sufficiently obvious, that a distinctive knowledge of diseases is of great practical importance in reference to their management. Treatment cannot be judiciously applied until a diagnosis has been reached. It may be based on the presence of characteristic signs peculiar to certain affections alone. Thus the crepitant sound denotes the existence of *pneumonia.* But there are very few signs which are inseparable from a disease, being found in that and no other. It must be remembered, that all the symptoms typical of a disease will be but rarely present.

In some, possibly many instances, the disorder is readily apparent, and the exact location of it detected. Cough, and rapid breathing would point at once to the chest as the seat of the trouble. In other cases much difficulty will be experienced. Some symptoms may be absent, and others unduly prominent. Patient watching, with a careful analysis and study of each individual case will, however, usually dissipate all doubts.

The previous history is essential in reaching a diagnosis. The duration of the symptoms materially assist in determining whether we have an acute or chronic disease to contend with. If the animal were recently to all appearances in good health, and the attack more or less sudden in its invasion, the disease is probably the former, whereas, if the dog has for a long time shown certain signs of ailing, the disease is probably chronic in character. A highly effective method of reaching the diagnosis of a disease is called, "reasoning by way of exclusion."

In a case of doubt, the problem is generally to decide between a certain number of diseases. The existing disease is one of two, three, or more, which may be suspected. Now if it be difficult to decide which one of these is the disease present from positive proof, it may be practicable to decide that there is insufficient evidence of the existence of one or more, and therefore they are excluded. By this process of elimination, the number of diseases is diminished, and may be reduced even to one disease. To illustrate the application of this method.—

Our dog does not respond to the usual call. We seek his kennel for the cause, find him unable to leave it, and observe the following symptoms. His breathing is rapid and labored; his manner exceedingly dull; he opens his eyes only to close them at once; hangs his head, it falls as though he slept, to be lifted as he seems to waken, or disturbed from time to time by a dry hacking cough, with an attempt to vomit, occasionally raising a little colored sputa. Nose and body are very hot. He lays down only to assume at once a sitting position, with forelegs braced and separated.

The disease may possibly be in the throat, but is evidently in the chest. The previous history tells us that the symptoms are *acute*, therefore we can at once eliminate all *chronic* affections, and there are left *laryngitis, bronchitis, asthma, pleurisy, and pneumonia.*

An examination of the throat dispels the doubt respecting *laryngitis.* No knowledge of a previous attack, the absence of wheezing respiration, and husky, barking cough, and *asthma* is reasonably excluded.

In *bronchitis*, so early in the disease, we should not expect such marked constitutional symptoms; while some fever would probably be observed, it would scarcely run so high; again, while the breathing is often accelerated, it lacks in the early stages at least, that labored character. Pain in *bronchitis* is evident when the patient coughs, and is less apparent in the interval. His discomfort would tend to make him restless, and on lying down he would assume no unusual position. These facts considered, render *bronchitis* improbable.

Uneasiness of the animal is one of the marked symptoms of *pleurisy.* That indication is absent in the patient before us. The breathing too is different, while in *pleurisy* it is labored, it is also unmistakably painful, and inspiration is shortened from that cause. A dog affected with that disease, would seem to avoid taking more air into his lungs than absolutely possible. There would be a restraint in the working of the muscles of the chest, that plainly told of pain. This too would be shown in the cough, dry and shortened, with little or nothing raised. While fever is present in *pleurisy*, it seldom in the early stage, runs as high as observed in this case.

If these differences in symptoms noted are insufficient, an examination of the chest by the ear will remove what doubts remain.

Pleurisy then excluded we come at last, by this method, to the disease of the animal before us — *pneumonia.*

PROGNOSIS.

The art of foretelling results in diseases is called prognosis. To determine the probable end in many cases, is often important with reference to treatment. The writer has deemed it wise to consider the signs on which prognosis is based, incidentally in connection with individual diseases. Some few obvious appearances which render the prognosis unfavorable, may here be mentioned: — *Continued loss of flesh* when connected with chronic affections is *serious. Very feeble, rapid pulse*, more especially in acute diseases, indicates a *very grave condition.* Among the *fatal signs* are *a fixedness of the eyes*, denoting paralysis; *involuntary discharges*, indicating *great insensibility.* A *jerking inspiration* if not dependent upon diseases of the lungs, *betokens death.*

CAUSATION.

A knowledge of the *causes* of disease is highly important as a means of *prevention*. When causes are traced to their origin, their influences can' often be obviated. It not unfrequently happens that ignorance of etiology largely prejudices recovery, when a knowledge of the morbid influences, still operative, might be removed.

The term *traumatic* is used to distinguish certain causes. Anything which occasions an injury or wound of a part, from which disease results, is a traumatic cause. Certain diseases, the origins of which we are unable to appreciate, are termed *spontaneous*.

Predisposing or exciting causes are influences that induce a tendency or liability to certain affections. They alone may be sufficient to give rise to disease, or they may only suffice to so influence the system, the conditions will be favorable for the occurrence of it.

ADMINISTERING MEDICINE.

It is presumed that the owner, or at least one familiar with the dog to be treated, will administer the medicine. If kindness and patience is exhibited, little or no trouble will be experienced.

An important object is to *concentrate drugs* as much as possible. If given in the form of small pills, they can be pressed into raw beef and thrown to the animal, after first tempting him with a few pieces.

When medicine is to be given in a bolus or very large pill, this method should be employed. Grasp the muzzle of the dog firmly with the left hand, the thumb and fore-finger on either side, pressing in the upper lips covering the teeth, thus preventing his biting. His mouth being opened, and head elevated, carry the bolus back into the throat as far as possible and close his jaws. If he does not swallow immediately, closing his nostrils, and stopping his breathing will be effectual.

If the dog is very large or unruly, and liquid medicine of unpleasant taste is to be administered, an assistant will be needed.

Liquids can best be given from a bottle, the assistant preventing the jaws closing and breaking it. A spoon is unfit, as much will be spilled in its use. Making a funnel of the cheek and pouring the medicine in slowly, is an admirable method, and can be employed if the dog is unconscious and cannot swallow; care must be observed, allowing but little at a time to trickle down the throat.

Some medicines can be disguised in milk or strong broths. Ordinarily' drugs should be given upon an empty stomach, excepting tonics, which should enter with the food, and be incorporated with it.

CHAPTER IV.

SPECIAL PATHOLOGY

AND

THERAPEUTICS.

INTRODUCTION.

Before entering upon the consideration of individual diseases, it may be well to understand certain points of distinction, interesting and of practical importance.

Differences as regards severity and duration, constitute a basis of the division of diseases into varieties. The same disease may be either *acute*, *subacute*, or *chronic*. A disease is acute when it has a certain degree of intensity, and runs a rapid course. The subacute variety has less intensity; a disease of moderate activity. The chronic variety exists when a disease is subacute, and it is of long continuance. The division into varieties based on the difference just named, is especially applicable to inflammatory affections.

A point of distinction of practical importance relates to *duration*. Some diseases continue for a definite period, never exceeding certain limits in this regard. Those which tend intrinsically to end after a certain time, are distinguished as *self limited*.

In the list of individual diseases, are some recognized as such for the sake of convenience, but which in reality, are only effects or symptoms of disease, as for example, *jaundice* and *dropsy*, which are merely symptoms occurring in connection with different affections.

It will be observed that the term expectant is often used. the treatment of a disease by expectation, consists in watching carefully its progress, and meeting, with appropriate measures, unfavorable events as they arise, or withholding active treatment until the need is manifest.

It will be noted that in prescribing drugs, simplicity has been observed. Not the least important is the selection of concentrated remedies, divesting them of nauseousness, thereby rendering their administration less difficult.

The writer in prescribing medicines, has deemed it wise and more convenient to choose some standard, and leave the reader to divide or add to the doses as the size of the dog may warrant. He has therefore selected one of *about fifty pounds* in weight. Where a *toy dog* is to be treated, the dose should be *lessened one half*, and for a *mastiff* or *St. Bernard*, or one approaching in size, the dose should *be doubled*. For instance, if the dose *one teaspoonful* is recommended, that would be proper for a *setter* or *pointer*, in fact any dog of fifty pounds weight; *one half a teaspoonful* would be the dose for a *terrier* or *pug*, and *two teaspoonfuls* for the *larger breeds*.

Sufficiently correct for ordinary purposes — *a teaspoon* measures one liquid drachm — *a dessertspoon* two drachms — *a tablespoon* four drachms, or one half an ounce.

An examination of a dog when ill cannot be hurriedly made. By patient watching some clue as to the location of the disease, may be acquired. His general appearance, the expression of his eyes, his breathing, his manner of moving about, should be noted. He should be forced to walk, and his back be particularly observed, if natural or arched. His way of lying down, and the position assumed may signify much.

By watching the head, the point of suffering may often be determined. The ears, mouth, tongue, and throat should be carefully examined. To note the pulse and temperature is of especial importance. The former, normally runs from eighty to one hundred. The indications from it may be read thus:— When irregularly intermittent— *nervous affections;* continuously intermittent — possibly *organic disease of the heart;* rapid and bounding; *fever, or inflammations ;* soft and easily compressible — *debility* and *depression;* thin and thready — *exhaustion and death.*

The examination of the abdomen should be carefully made. Observing if the same is soft and flaccid, or hard and tense. Tenderness beneath the hand, will be revealed on gentle pressure, by the animal shrinking, or turning with piteous expression.

As regards *nursing*, pure air, sunlight, cleanliness and warmth, nourishing and sustaining diet, are the essentials.

In sickness and in health alike, man's true friend and honest comrade the dog, deserves all the care his owner can bestow upon him.

DISEASES

OF

THE RESPIRATORY ORGANS.

ACUTE PLEURISY.

Each lung is invested, upon its external surface, by an exceedingly delicate membrane, the pleura, which incloses the organ as far as its root, and is then returned upon the inner surface of the chest. The inner surface of the pleura is smooth, polished, and moistened by a fluid which favors the easy play of the surfaces, as the lungs alternately expand and collapse in movements of respiration.

In inflammation of this membrane or pleurisy, this lubricating fluid disappears, and the surfaces become roughened, hot, swollen, and painful as they come in contact and rub together with every inspiration. This condition rarely exists longer than twenty-four hours, and oftener a much shorter time; then follows an effusion into the affected side, the space being in some instances only partially filled, and in others the amount of fluid is of sufficient quantity to fill the cavity of the pleura, compressing the lung into a small, solid mass. In certain cases of acute pleurisy which pursue a favorable course, absorption of the liquid commences a few days after the accumulation has reached the maximum. Should the quantity remain stationary, or the diminution take place very slowly, after the elapse of two or three weeks, the disease becomes chronic.

Pleurisy is an unilateral disease, that is, it affects the pleura of one side only. While it may be associated with pneumonia (pleuro pneumonia), it does not tend to the development of that disease.

Causation.—Acute pleurisy may be produced by contusions, especially if accompanied with fracture of the ribs, and penetrating wounds. It is remarkable, however, that severe injuries of the chest often occur without giving rise to this disease. Cold resulting from exposure, is the more frequent source of pleurisy. It may, however, be spontaneous, that is proceed from an unknown internal cause.

Symptoms.—The attack is usually sudden. In a certain proportion of cases, however, some pain or soreness exists one, two, or three days before the development of acute inflammation. It is sometimes ushered in with a chill, as shown by shivering. Pain in the affected side attends the onset in a majority of cases, and is usually intense. It is sharp and cutting in character and is felt especially in the act of inhaling. It increases during inspiration, often becoming so severe that the act is shortened and arrested before completed. This is due to the pain, caused by the rubbing together of the inflamed sides of the pleura as the lung expands. The respiration is consequently quickened, and the animal instinctively multiplies the acts to compensate for the want of a full inspiration.

Cough is usually present. The pain in coughing leads instinctively to efforts to repress it, and its character is termed suppressed.

Fever is coincident with the development of the inflammation, varying in intensity in different cases.

The usual indications associated with fever are present; restlessness, thirst, pulse rapid, full, and bounding, nose hot and dry, tongue slightly coated, eyes watery, and wearing a pleading anxious look, the whites reddened and the lining of the lids deeper in color.

After a considerable amount of effusion has taken place, the symptoms are materially modified. The pain is notably lessened, the acts of coughing are less distressing, and the efforts at suppression not so apparent. The fever diminishes, and the other evidences of constitutional disturbance, in a manner disappear. The respirations continue, more or less hurried, their frequency now depending on the compression of the lung by fluid.

If a considerable amount of liquid is rapidly effused, the respirations are quickened; the animal suffers from a painful sense of the want of breath, and may be obliged to maintain a sitting posture, with fore legs spread widely apart. If the quantity is not large and the effusion has not taken place rapidly, less suffering will be noted while quiet, but exertion will cause panting and a sense of suffocation.

With a small quantity of effusion, the animal prefers, when lying down, to lie on the affected side, as in that position he is able to expand more fully the opposite lung. With a large quantity of fluid present, the breathing becomes more labored and abdominal as shown by the heaving action of the muscles of the flank. The limbs become dropsical, suffocation is threatened, and death may result from that cause.

Diagnosis. — In the earlier stages some difficulty will be experienced in discriminating between the disease under consideration, and pneumonia. An examination of the chest by inspection, will show restrained movements, caused by the pain. Examination by the ear will determine the respiratory sound more or less weakened on the affected side, due to the fact that pain leads instinctively to a diminished use of the lung involved, while the action of the other lung is increased. No great importance can be attached to this sign alone, as in some attacks of pneumonia, severe pain may exist, and the same appearance be present. A pleural friction sound is a sound of grazing, rubbing, or grating, due to the movements, in opposite directions of the pleural surfaces with inspiration and expiration. The sound is more or less intense, dry, and appears to be near the ear, conveying the idea of friction of roughened surfaces. If this is distinctly present, its diagnostic significance is important, showing pleurisy exists.

The signs belonging to the second stage manifest themselves without much delay. Liquid effusion having taken place in sufficient quantity to be apparent, the diagnosis is easier.

The signs of an effusion are obtained by percussion, auscultation, and inspection. To determine by percussion, press the palm and fingers of the left hand firmly against the side of the chest, then tap lightly one finger with the second finger of the right hand, giving a sharp quick blow. If the sound given forth is dead and flat, it denotes an absence of air within the part of the chest percussed. On auscultation the respiratory sound is either suppressed or but feebly heard, unless the quantity of liquid is sufficient to compress the lung into a solid mass, then the sound is intensified, rude, and harsh, like that heard on placing the ear to the neck of the animal. If the quantity of liquid be large enough to distend the pleural sac, the depression between the ribs of affected side may be diminished and even swell out. This complete distension by liquid effusions is, however, exceptionally rare.

Prognosis.—The prognosis in acute, uncomplicated cases of pleurisy, with but little effusion, is favorable, provided the constitution of the animal is not enfeebled. The presence of a large liquid effusion is a very unfavorable sign, and death may result from suffocation or exhaustion.

When recovery takes place from pleurisy, the roughened surfaces of the pleura come together and are united by adhesion of newly found tissue.

Treatment.—The objects of treatment in acute pleurisy, differ in its different stages. The purposes in the first, are to arrest, if possible, the progress of the disease, to diminish its intensity, if it be not arrested, and to relieve suffering. The indications are essentially the same as in the earlier period of most acute inflammations.

The remedy of the first importance is opium in some form. It is invaluable not only for its palliative, but its curative action. By relieving pain, it diminishes the determination of blood to the inflamed parts. It also lessens the constitutional disturbance, and enables the system to tolerate better the local affection. Conjoined with the use of opium, remedies addressed to the circulation, may be employed. Aconite is preferred, because its sedative effect upon circulation is produced without depression. Blisters should not be applied in this stage. Cloths wrung out of hot water and bound around the chest will act as a palliative, and afford marked relief.

Opium and aconite are conveniently combined in the following mixture—

> R. Morphiæ Sulphatis   gr. ij
>
> Tr. Aconiti Rad.   gtt. xxv
>
> Aquæ   ℥ iij
>
> Ft. Mist. Sig. Dose one teaspoonful.

This should be given every three or five hours until the pain subsides and the animal is inclined to sleep.

While the pain continues, and the fever remains unabated, the sedative mixture should be persisted in. A mild purgative, in the form of syrup of buckthorn in tablespoonful doses may be added to the treatment, to aid in lessening the fever, and restraining the liquid effusion.

In the second stage after the pain and fever have subsided, the object of treatment is to promote absorption of the liquid. The kidneys may be stimulated to increased activity by the following—

> R. Potass. Acet.   ℥ iv
>
> Infus. Digitalis   ℥ iv
>
> Aquæ   ℥ ij
>
> Ft. Mist. Sig. Dose two teaspoonfuls every five or six hours.

Counter irritation may now be resorted to, the tincture of iodine applied with a brush preferred. The important object is to support and nourish the animal. Laxatives and external applications must be discontinued if the general strength seems impaired by their use. The diet should be nutritious and concentrated. Tonic remedies are useful and may be given in the form of the tincture of iron, twenty drops well diluted, or quinine in two grain doses, three times a day. Throughout the disease, pure air, warmth, and cleanliness are absolutely essential.

During convalescence, the object is to re-establish the normal health. Gentle exercise out of doors is to be encouraged, and the local affection will disappear in proportion to the rapidity with which improvement in the general condition takes place.

CHRONIC PLEURISY.

Chronic pleurisy occasionally follows the acute form of the disease, but in the majority of cases is a sub-acute affection from the first. It is often developed imperceptibly, those prominent symptoms observed in the acute form, being absent or lacking sufficient intensity to characterize the disease.

Anatomically, the conditions existing in chronic pleurisy are essentially the same as in the acute form. The pleural cavity contains a liquid effusion of serum and lymph, which in the event of recovery becomes absorbed. Bands of adhesion are formed uniting the inner surface of the pleura, and the chest walls become contracted, leaving a depression generally permanent, more especially in cases of large effusions.

Causation.—The same influences may give rise to chronic that are involved in the causation of acute pleurisy. It may be induced traumatically, and is sometimes due to exposure. More often it proceeds from internal causes which are not understood.

Symptoms.—Cough, pain and tenderness are frequently wanting, and rarely prominent in this disease. The most marked symptom is the increased frequency of the respirations, which may be moderate while at rest, but become evident on exercise. Here, as in the acute form, the embarrassment in breathing will depend upon the amount and rapidity with which the effusion takes place. The pulse generally runs above the normal, and is small and compressible. Appetite is impaired.

Diarrhœa occurs in some cases, and emaciation may follow, although the strength is often maintained to a marked degree, and the animal be out and take considerable easy exercise.

Diagnosis.—The symptoms apparent in this disease will aid much, but a positive diagnosis can hardly be reached without auscultation and percussion. The signs of liquid effusion are the same in chronic as in acute pleurisy. In the former the amount is often greater, and the contraction of the affected side is more marked after absorption.

Chronic pleurisy is often associated with other diseases, among them pulmonary consumption and dropsical affections, which, by their individual symptoms, may be detected.

Prognosis.—Simple chronic pleurisy unassociated with any other serious disease, and occurring in an animal previously healthy, may be recovered from, but a permanent deformity, in contraction of the affected side is quite certain to result. Death often occurs from associated complications. In many cases a breaking down of the lungs follows chronic pleurisy, and the disease assumes the form of consumption. In other cases absorption of the effusion may go on to a certain extent, and the animal regain good health, even although a portion of the liquid remains.

Chronic pleurisy may end in empyema, in which disease the liquid becomes purulent, and from which recovery is very doubtful.

Treatment.—The objects of treatment are to remove the effused liquid, and develop and sustain the general strength of the system. For the first object, the measures which may be employed, are the same as in the last stage of acute pleurisy. Small blisters, or the tincture of iodine may be applied to the affected side, and mild laxatives should be judiciously given if the general condition and strength warrant their use.

The second object calls for tonic remedies, which should be given three times a day. A pill, combining several of the more powerful agents, can be made as follows —

> R Quiniæ Sulph.   gr. xxiv
> Ferri Carb.   ℥ ss
> Ext. Nucis Vomicæ   gr. viij
> Ft. Pil. No. xxiv. Sig. Dose one three times daily.

Tonics should be persisted in, and may be changed occasionally, as the animal becomes accustomed to their use. The pills may, after a time, be discontinued, and the syrup of the iodide of iron in ten drop doses, or the elixir of calisaya bark, iron, and strychnia in teaspoonful doses, be given. If the animal be thin in flesh, cod liver oil must be added to the treatment.

Pure air, gentle exercise, and nutritious diet are essential in this, as in all exhausting diseases.

In extreme cases, where a rapid formation of the liquid effusion threatens death by suffocation, tapping the chest may be resorted to, but a skilful physician, the family attendant, should be employed for the operation, which, in but few cases, will result unsuccessfully, as the fluid generally returns.

PNEUMONIA.

An inflammation of the substance of the lungs, or pneumonia is characterized by an exudation into the air cells, solidifying for a time that portion of the lung affected. The first appreciable anatomical change in this
disease is an abnormal accumulation of blood, due to an active congestion
of the parts involved. This condition lasts but for a short time, and is
speedily followed by an exudation, a material which escapes from the
blood and coagulates within the air-cells. The cells are filled and distended with this exuded matter and cease to contain air. The lung is
solidified, presenting an appearance not unlike liver, and hence this condition is called hepatization.

If the progress of the disease be favorable, the exudation is removed
mainly or exclusively by absorption, and the recovery of the affected
part is complete. If the disease progresses unfavorably, absorption of the
product within the air cells does not take place, the substance of the lung
softens and breaks down, a condition called purulent infiltration.

The constitutional symptoms then become more marked, the distress
more apparent. The expression grows even more piteous and haggard,
the eyes more sunken. The nostrils dilate, the cough loose becomes persistent and distressing, and the mouth and lips are covered with a thick
slime. The breath which has a sweet, sickish odor, comes quick and short.
Approaching suffocation is obvious. The sitting position so long maintained is still persevered in, until the limbs become dropsical, the skin
cold and clammy, the pulse faint and flickering, finally death ends all.

Causation. — While exposure to cold, is an evident cause of pneumonia
in many cases, it is probably but an exciting influence co-operating with
an existing morbid condition or predisposition, which we are unable to
comprehend. The disease is in many cases developed spontaneously, no
obvious cause apparent.

Symptoms. — The invasion of the disease is often abrupt. In a large
proportion of cases pneumonia is ushered in with a chill, as shown by
shiverings. Speedily following the chill pain is manifest, in some cases
severe, in others very slight. Cough usually appears soon after the attack
commences. It may be prominent, and is sometimes wanting. The matter
coughed up is at first scanty and thick. In a certain proportion of cases
it soon assumes a character distinctive of the disease termed rusty, from
its reddish tint like iron rust.

Fever follows immediately the invasion of the disease. The pulse runs
high, loss of appetite occurs, great thirst is experienced, and prostration
manifested. The nose becomes hot and dry, the eyes reddened. The respirations increase in frequency, as the disease progresses, until a sense

of suffocation forces the animal to assume a sitting position, with extend-ed head and protruded tongue.

Diagnosis. —During the early stage, by placing the ear to the affected side, crepitation can be heard in most cases. This when present is highly distinctive of pneumonia. The crepitant sound is dry, very fine, crackling, and heard only on inspiration. This sound is due to the separation or bursting of agglutinized bubbles, and is much the same as one hears made by the boot when walking on crusted snow.

In the second stage, after the lung has become solidified, percussion shows marked dulness, a flat dead sound being emitted. The respiratory sound becomes exaggerated by the law of transmission of sound by solids, and the bronchial respiration is distinctly heard. Crepitation may be present in this stage, and may be wanting.

When recovery from the disease commences, the respiratory sound grows more natural, crepitation at times returns, and frequently a sub-crepitant sound is heard. This is fine, moist, and bubbling, carrying the idea of small bubbles, heard with either inspiration or expiration, or with both acts.

Dulness on percussion, becomes less and less marked, and finally dis-appears, but a trace is often left for a time after an examination by the ear shows the exudation to have been removed. If the disease does not clear up, but passes into the stage of infiltration, the dulness or flatness on percussion continues, and coarse bubbling sounds are detected, due to the presence of pus in the air-cells, heard at times louder just after cough-ing, and can often be appreciated by placing the flat of the hand to the chest, when a sensation of fine bubbling will be transmitted.

Prognosis. —Pneumonia is a very serious disease, and is very often fatal. At times only a small proportion of the lung is affected, the disease then being termed circumscribed. The extent of the disease in a great measure indicates the gravity and danger. Accidents and complications are very liable to invest the cases with greater gravity, and where there are no symptoms which denote imminent danger, sudden changes often take place for the worse.

When convalescence in the disease sets in, recovery is generally com-plete. From purulent infiltration there is very little hope of recovery.

Treatment. —Pneumonia belongs among those diseases distinguished as self limited. If uncomplicated and not attended with accidents, it runs a definite course, ending in restoration, provided the powers are main-tained. The treatment must be supportive and every influence calculated to depress the system, must be studiously avoided.

In the first stage, the object should be to relieve symptoms, to diminish the intensity of the inflammations, and prepare the system to bear the severe strain forced upon it. Many authors advise the use of blisters, but this the writer cannot endorse, considering their employment injudicious.

Other authorities recommend flaxseed meal poultices, and warm fomentations. These are objectionable from the fact that the animal is liable to become chilled while the applications are being removed and changed.

The object is to prevent the air coming in contact with the chest, and no better means can be devised than the cotton jacket. Many thicknesses of cotton batting can be quilted on to cloth, cut with holes for the fore legs, and roughly shaped to envelop the chest from the neck to the loins. This when basted on to the animal is a sure protection, and contributes to his comfort, by keeping the skin moistened with perspiration, and all the advantages of a poultice are thereby secured.

Opium should be given in the first stage to subdue the pain and to quiet the system. If the pulse is full and bounding and the animal in good health previous to the attack, sedatives may be combined with the opiate, and the morphine and aconite prescription recommended in acute pleurisy be used. If pain is not severe, and fever runs high, the tincture of aconite root may be given every two hours, alone, or in the following mixture —

R	Tr. Opii. Camphoratæ	℥ i
	Tr. Aconiti Rad.	gtt. xxv
	Potass. Chloratis	℥ ij
	Ammon. Muriatis	℥ i
	Syr. Simplicis	℥ i
	Aquæ	ad. ℥ iij

Ft. Mist. Shake well, Sig. Dose one teaspoonful from four to six hours.

A room of equitable temperature should be chosen, and good ventilation be maintained. Never wait in pneumonia until trouble comes; rather anticipate it, and be prepared for emergencies. Commence early with supportive measures, and insist upon it the animal takes nourishment at regular intervals. Select the concentrated and most sustaining nutriment such as raw eggs, beef juice, and beef extracts.

To support and sustain the vital powers and to prolong life beyond the limit of the disease, is the general indication in the second stage. Opium if indicated may still be persisted in, and stimulants and stimulating tonics should be commenced. Given moderately at first, but pushed if the need becomes manifest. Quinine in two grain doses should be given every four hours, and two teaspoonfuls of brandy may be administered in milk at intervals, and the effects watched. Should the pulse weaken, the stimulant must be increased to a tablespoonful dose, and given every two or three hours. If failure of the powers seems imminent, one teaspoonful of the aromatic spirit of ammonia must be added to each dose of brandy.

If the limit oi the disease is reached, and resolution commenced, recovery promises to be rapid, and there is little or no danger of a renewed attack. In the stage of resolution, tonic remedies are indicated, the quinine can be discontinued, and some combination of iron be substituted. The elixir calisaya bark, iron and strychnine, in teaspoonful doses before feeding, is an admirable tonic. It is sometimes difficult to secure a reliable preparation however, in which event the following mixture will be found useful—

    R Ferri et. Ammoniæ Cit.   ʒiss.
        Tr. Nucis Vomicæ   ʒij
        Tr. Cinchonæ Comp.   ʒiv

Ft. Mist. Sig. Dose one teaspoonful three times daily before feeding.

Cod liver oil should be added to the treatment, and given in tablespoonful doses after eating or mixed with the food. As soon as the animal is fairly convalescent a solid diet may be allowed, meat raw or cooked as he seems to prefer, in fact anything within reason that he will eat, and the quantity need not be much restricted. Exercise in the open air in pleasant weather should be early encouraged.

In event the lung breaks down, and the disease passes into the stage of purulent infiltration, there is little or no hope for the animal. The only treatment is to crowd the stimulants and nourishment.

Recent authorities have advised the use of chloral hydrate, to combat the pain in pneumonia. The writer feels that there are decided objections to its use, from its peculiar depressing action on the heart, which organ throughout the disease needs to be stimulated and strengthened.

BRONCHITIS.

An inflammation seated in the lining membrane of the larger bronchial tubes constitutes the disease called bronchitis, a very common affection among dogs. It is generally ushered in by an inflammation of the mucous membrane of the nasal passages, which commencing in the nostrils travels downward to the lungs.

The disease may be divided into two stages, which can be conveniently termed the dry and the moist. From the occurrence of the first symptoms in the nose, in from one to three days the disease makes the transit to the larger bronchial tubes, the mucous membrane of which becomes dry, reddened, swollen, and sensitive. In the progress of the affection, after two or three days, the mucous secretion, which has been absent or scanty, reappears and is abnormally increased.

Bronchitis may be primary or secondary, that is it may occur as a distinct affection or be developed as a complication of certain diseases, such as pneumonia, or consumption.

It may be either acute, sub-acute, or chronic, as based on the differences as regards the severity and duration of the inflammation.

Causation.—Exposure to damp and cold; violent exercise, and subsequent confinement in draughty kennels; changes in the weather, and some special atmospheric influence not understood. This is to be inferred when the disease prevails as an epidemic.

Symptoms.—Shivering, and the common evidences of catarrh, are the primary symptoms noted, followed by some fever, higher pulse, thirst, loss of appetite, dulness and general lassitude. The cough is at first dry and painful, but not suppressed. This changes as the mucous secretion becomes abundant, then the pain abates, the cough loosens and has a softer tone. Respiration is usually unaffected, rapid breathing being observed only in exceptional cases, in which the mucous secretions accumulate, and obstruct the bronchial tubes. The expectoration is at first small and scanty, later copious and frothy. The eyes are reddened, the nose hot, and glued with a thick mucous which becoming free and thin, is accompanied with frequent sneezing.

These symptoms are present in ordinary bronchitis. An unfrequent and graver form is developed when the inflammation affects the smaller tubes. This is termed capillary bronchitis, and differs widely from the ordinary acute disease. It is an exceedingly grave affection, and the danger proceeds from obstruction to the current of air to and from the air-cells. This obstruction is due to the small size of the affected tubes. The swelling of the mucous membrane, and the presence of the muco-purulent liquid, which in the larger tubes do not interrupt the free passage of air during respiration, here occasions serious interference. In the capillary form of bronchitis, all the symptoms of suffocation are present and death usually results from that cause.

Diagnosis. — Acute bronchitis is to be discriminated from pleurisy and pneumonia, more especially the latter. The absence of severe pain and rapid breathing, the presence of a copious expectoration, which is not rusty, are negative points, which, with an examination of the chest, will render a diagnosis easy.

On percussion nothing abnormal is detected. On auscultation coarse mucous rattles are heard throughout the chest. Their presence and diffusion over both sides are distinctive of the disease. The gravity of the symptoms will indicate whether capillary or simple acute bronchitis is present.

Prognosis. — Ordinary bronchitis when unassociated with other diseases is dangerous only in very young puppies and old dogs. A mild attack is soon recovered from. In the capillary form, there is little or no hope for the animal so affected.

Treatment. — A very important item is to guard against changes in the temperature. The patient should be kept in a room well ventilated and comfortably heated. In the earlier stage, an attempt should be made to abort the disease, with from five to ten grains of Dover's powder at night, and an application of mustard and vinegar, or kerosene oil to the chest, anointing the parts afterwards with fresh lard. This treatment should be followed the next morning with a dose of castor oil or syrup of buckthorn.

The effort if unsuccessful will at least lessen the severity of the disease. For the cough in the earlier stage the following is recommended —

> ℞ Syr. Scillæ   ℥i
> Spts. Æth. Comp.   ℥i
> Syr. Tolu   ℥ij

Ft. Mist. Sig. Dose one teaspoonful every three or four hours.

If the cough becomes a distressing symptom, opium can be wisely combined with expectorant remedies, as in the following—

> ℞ Sol. Morph. Sulph.
> Syr. Scillæ
> Vini Ipecac
> Syr. Senegæ   aa ℥i

Ft. Mist. Sig. Dose one teaspoonful every three or four hours.

In the second stage, nutritious diet and tonic remedies are measures which assist in recovery; two grains of quinine may wisely be given two or three times a day.

The chlorate of potassa is a remedy of great value in all diseases of the mucous membranes, more especially when the secretions are scanty. It will be found efficacious in all stages of bronchitis, and may be added freely to the water the animal drinks, or given in the following mixture—

R   Potass. Chloratis $\mathfrak{Z}$ ij
   Syr. Scillæ $\mathfrak{Z}$ ss
   Syr. Tolu. $\mathfrak{Z}$ ij
   Aquæ   ad. $\mathfrak{Z}$ iv

Ft. Mist. Sig. Dose one teaspoonful every three or four hours.

If the disease involves the smaller bronchial tubes, and suffocation is threatened, an emetic should be administered and a stimulating expectorant given as in the following—

R   Ammoniæ Carb. $\mathfrak{Z}$ ij
   Syr. Senegæ $\mathfrak{Z}$ iij

Ft. Mist. Sig. Dose one half a teaspoonful in a tablespoonful of peppermint water every two hours.

When danger of sinking is evident, whiskey or brandy may be added to the treatment and given in dessertspoonful doses, as often as the urgency demands.

If the disease tends to pass into the chronic form, cod liver oil two or three times a day should be given, with an iron tonic of elixir calisaya bark, iron, and bismuth, dose one teaspoonful.

As a rule expectorant remedies are not indicated in chronic bronchitis, and do harm by their depressing effect, and by disturbing the appetite and digestion.

CONSUMPTION.

The term consumption, wasting of the body, strictly speaking, implies pulmonary tuberculosis, an affection characterized by the presence in the lungs of a morbid product called tubercle. This deposit eventually undergoes softening, becomes liquefied, ulcerations follow, and destruction of the lungs result.

Causation. — Every influence which appears to affect the health has been alleged to be a cause of consumption, by impairing the nourishment of the body and inducing poverty of the blood. Bad air, deficiency and poor quantity of food are prominent factors. Poor food injures the mucous membrane of the stomach and intestines, and produces repeated attacks of irritation in those parts. That the seeds of consumption are planted by breeding in and in, is quite probable.

Antecedent diseases of the respiratory organs may induce the affection directly, and other serious derangements tend to its development. Jacobi reports that a dog which ate the sputa of his master, ill with the disease, died of consumption; an evidence of communicability.

Some authors include asthma among the causes of consumption. This is not the case however, as the presence of asthma seems to afford protection against pulmonary tuberculosis.

Symptoms. — In many cases, latent symptoms exist without sufficient prominence to excite suspicion, and a cough dates the beginning of the disease. The number of respirations is somewhat increased, the heart's action irritable, and the pulse accelerated. The loss of flesh is perceptible, the appetite somewhat diminished, the coat becomes staring and the hair falls out. Fever sets in and runs at intervals. The strength is obviously impaired. Emaciation gradually becomes extreme, diarrhœa sets in, the animal becomes exhausted and dies. The symptoms and course of the disease are modified by complications, or as other organs are involved.

Treatment. — In the earlier stages certain agents may be employed with a hope of arresting the disease; the results will however in the majority of cases prove negative. The compound syrup hypophosphites may be given in teaspoonful doses before feeding, and cod liver oil in tablespoonful doses two or three times daily with the food. The diet should be nutritious and concentrated. Pain and distress demand the administration of opiates.

ASTHMA.

The term asthma should properly be limited to one morbid condition. Veterinarians distinguish between the congestive and spasmodic forms. The congestive and more common affection observed among dogs, is due to an inflamed state of the mucous membrane of the bronchial tubes. In a great majority of cases bronchitis exists, and the susceptibility to the exciting causes is increased by its presence. The disease is more common in old age, and among obese, over fed, house dogs. The predisposing causes are derangement of the circulatory system, such as heart and lung diseases, and chronic indigestion.

Causation. — The existing causes are many; cold from exposure, acute bronchial inflammation, and certain inhalations are particularly liable to bring on an attack.

Symptoms. — Labored efforts of breathing prompted by a sense of suffocation, croupy respirations not increased in rapidity, and at times less in number than normal. Inspiration jerking, expiration accompanied with a wheezing sound. Distress and greater difficulty in breathing on exertion. Nostrils dilating, expression piteous and anxious. Cough in many instances exists prior to, or manifests itself during an attack of asthma. A dilatation of the air cells is generally associated with this disease if severe and of long standing.

Diagnosis. — The diagnosis is not difficult; the disease has such distinctive differences, it need not be confounded with other affections in which labored breathing is a prominent symptom.

Prognosis. — The significance of the disease depends upon the causes and conditions which induce it. In itself, asthma is not an affection likely to prove fatal, but it must necessarily shorten life.

Treatment. — During an attack a prompt emetic of the sulphate of zinc twenty grains, or ipecac thirty grains, will afford relief. This should be followed with a cathartic of syrup buckthorn and castor oil, of each one half a tablespoonful. If the violence of the attack does not subside, the animal should be shut into a closed room, and on a pan of coals, stramonium leaves be burned, to an ounce of which, a drachm of nitrate of potassa has been added. This measure will relax the spasm, and induce expectoration. To relieve the co-existing bronchitis, should next be attempted, and the following given —

℞	Potass. Chloratis	ʒij
	Tinct. Belladonnæ	ʒiss
	Tinct. Cinchonæ Comp.	ʒij
	Aquæ	ʒij

Ft. Mist. Sig. Dose a dessertspoonful four times a day.

If the animal is too fat and gross, exercise should be insisted upon, the diet simplified, and a laxative be given daily unless the condition of the bowels contra-indicate. The following pill will be found efficacious.—

℞   Ext. Aloes Aq.   gr. xv
     Ext. Colocynth. Comp.   gr. xv
     Podophyllin   gr. iv
     Ext. Nucis Vomicæ   gr. iv
Ft. Pil. No xij. Sig. Dose one daily.

In the treatment of asthma the laws of hygiene should be observed in kennelling and all exciting causes removed.

A radical cure is doubtful, but much can be done to mitigate the disease. With a possible chance of success, the iodide of potassium should be faithfully tried in all cases, and persevered in for some weeks. It may be given in the following mixture. —

℞   Potass. Iodidi   ℨ ij
     Spts. Ammoniæ Arom.   ℨ i
     Tinct. Quassiæ
     Aquæ   aa ℨ ij
Ft. Mist. Sig. Dose one teaspoonful three times daily.

In the purely spasmodic form of asthma, the burning of stramonium and saltpetre will be found especially efficacious, and the fluid extract of grindelia robusta in one half teaspoonful doses may be given every hour until relief is manifested.

ACUTE LARYNGITIS.

The larynx is the organ of voice, placed at the upper part of the air passage. It performs a double function, one part of which is connected with the voice, the other with respiration. It is situated between the base of the tongue and the windpipe at the upper and fore part of the neck.

The general character of laryngitis varies according to the extent and severity of the inflammation. The disease may run a mild uncomplicated course or it may be accompanied with an exudation of lymph, a condition distinguished as croup.

Causation.—Faulty kennel arrangements and consequent exposure to cold and wet. Extremes in temperature so often experienced by petted house dogs, which are by their habits, particularly predisposed to inflammations of the air passages. Mechanical irritations from prolonged barking and from foreign bodies not at once removed. Injuries to the neck, such as continued tugging at a chain may be sufficient to induce the disease. Inflammations of the adjacent parts, as is frequently the case in catarrh of the nasal mucous membranes, extend to the throat, involving the larynx.

Symptoms.—A hoarse, barking cough, excited by external pressure over the windpipe, is one of the diagnostic characters of acute laryngitis. In the earlier stage the expectoration is scanty and of thick mucous, later it becomes abundant, and is more easily expelled.

Fever may be more or less intense, or it may be wanting. If the swelling of the membrane be not great, the breathing is not notably increased. With a greater amount of swelling, labored respiration and difficult breathing are added. Inspiration and expiration are both affected by the obstruction, but more especially the former. It is harsh or noisy.

There is frequently more or less difficulty in swallowing, and a light, bubbling discharge from the nose. In cases where great obstruction exists the suffering is intense. The eyes are congested and expressive of the great distress the animal is experiencing. The suffering is at times rendered more violent by spasm, and the respiratory acts are less frequent than normal. With these symptoms, the disease in a large proportion of cases will end fatally, and the mode of dying be by suffocation.

Diagnosis.—The diminished frequency of the respiratory acts, and the absence, on auscultation and percussion, of signs denoting pulmonary disease. The peculiar husky character of the cough. The suffering increased spasmodically, with intervals of comparative ease. The tenderness over the larynx, and the reddened swollen appearance of the mucous membrane of the throat. The presence of these symptoms exclude all other diseases which might be confounded, and render a diagnosis easy.

Prognosis.—The prognosis depends essentially upon the severity of symptoms and efficiency and promptness of the treatment. When there is little or no infiltration of the sub-mucous tissues as inferred by the comparative ease in breathing, the chances are favorable. If the obstruction aside from that due to spasm, be sufficient to interfere greatly with respiration, a fatal result may be anticipated.

Treatment.—The injurious influences of changes in the air, should be avoided by maintaining a uniform temperature in the room in which the animal is placed. A sponge should be wrung out in hot water, and applied to the neck, being frequently changed. Five grain doses of Dover's powder will greatly relieve the discomfort, and in the milder cases its use is admissible. Opiates, in the severer forms of the disease, should be given guardedly, and not carried so far as to blunt the perception of the want of breath. A full dose of castor oil should be given early in the affection, to draw the blood to the intestines. Spasms in breathing are often relieved by emetics, and twenty grains of ipecac can be wisely given during an attack. The oil of copaiba is admirable in its action, and when inflammation of the larynx first manifests itself, doses of fifteen drops may be given from three to six hours as the severity of the symptoms indicate. It will be well to emulsify it in this way; add a teaspoonful of the oil to a raw egg, and beat well with a fork; give one fourth of the quantity at each dose. The following mixture is also very efficacious in this disease.—

R	Potass. Chloratis	ℨi
	Ammon. Mur.	ℨi
	Syr. Tolu.	℥i
	Aquæ	℥ij

Ft. Mist. Sig. Dose one teaspoonful in mild cases, every two hours. In very severe attacks, the same dose every fifteen minutes.'

If the disease assumes a severity threatening suffocation, the dog should be made to inhale steam or medicated vapors. If the room be too large to moisten the entire atmosphere, a tent can be made by placing a sheet over two chairs, under that the dog be placed, and a hot iron or stone be thrown into a pan of water. To medicate the steam add a tablespoonful of the compound tincture of benzoin for every quart of water. When death from suffocation is imminent and all other means prove futile, tracheotomy is the only hope remaining.

CHRONIC LARYNGITIS.

Chronic inflammation of the larynx often occurs without having been preceded by the acute form of the disease. It more commonly begins either as an acute catarrh, which from exposure and neglect assumes a chronic character, or in an insidious manner with insignificant symptoms at first, it by degrees develops more marked disturbances and eventually reaches a severity even in some cases affecting respiration. Hoarseness and a dry husky cough, which manifest increased severity at times, are among the characteristic symptoms.

The expectoration may be slight or more or less abundant; usually it is whitish gray, and frothy. The affection in many cases occurs in connection with consumption, and much the same treatment is demanded as in that disease. Cod liver oil should be given two or three times a day in table-spoonful doses, and if no gain is observed the compound syrup of hypophosphites in teaspoonful doses three times daily should be tried. The iodide of potassium in cases where no pulmonary disease exists, is an agent of value and may be given in the following mixture. —

> R Potassii Iodidi   ℥ ij
> Ammon. Carbonatis   ℥ ij
> Tinct. Hyoscyami   ℥ i
> Aquæ Camphoræ   ℥ iv

Ft. Mist. Sig. Dose one teaspoonful three times daily in a little water.

It must be remembered that no immediate effect from the iodide of potassium is ever experienced in any disease. its efficacy depends upon its prolonged administration. In all inflammations of the mucous membrane where the natural secretion is lessened, no agent has a more marked effect than the chlorate of potassa. An ounce may be diluted in a pint of water and a tablespoonful of the mixture be given every two hours, until the cough loses its dry husky character and becomes moist.

External counter-irritation to the neck by means of blisters and mercurial ointments, is of doubtful efficacy. Painting with the tincture of iodine daily instead, will have a happier effect.

A perfect and permanent cure of chronic laryngitis is a rare occurrence. The prognosis will of course be influenced by the associations or complications with other affections.

ACUTE CATARRH.

To use the term catarrh alone, as indicative of an individual affection seems unwarrantable, from its intimate attachments to a whole family of diseases, distinctively classified.

The literal meaning of the word is "to flow down," and originated in the idea the ancients had with regard to that secretion especially which escapes from the nose. It was believed by them that the secretion flowed down out of the brain, and the theory existed until Schneider, Professor in Wittenberg, published in 1660 his treatise "De Catarrhis," showing the falsity of this supposition. The term catarrh denoting inflammation of the mucous structure accompanied by an abundant secretion of mucous, has been used superadded to the name of the organ or parts involved. As the plan of distinguishing inflammatory affections by the suffix "itis" has become general, it is superfluous to use the terms bronchial catarrh, intestinal catarrh etc., when bronchitis and enteritis express as much and identically the same disease.

Acute catarrh or in plain language "a cold," is a frequent complaint among dogs and often engenders the most serious diseases.

Causation.—Physiologically the practice of "catching cold" is yet to be explained. This fact has been noted as the result of experiments, that when animals are subjected to a high degree of heat and then suddenly changed to the ordinary temperature, the heat of the body sinks, not to the normal degree but below it, the cause being accepted that the blood vessels are paralyzed for a time by the high temperature. It will be seen by this that dogs which are in winter allowed to remain in the house near hot stoves, and then to go out into the cold air, show a decided predisposition to the disease.

Allowing a dog a hard run and then shutting him into a cold stable or kennel, his body over-heated and perspiring, it will be appreciated, is a fruitful source of the disease. Damp, draughty kennels, and exposure in rain storms, and similar injurious influences are often immediately followed by the onset of the disease under consideration. The exertion of these influences is materially greatest at times when the weather is cold and moist. It is easy to understand that when these causes extensively exist, the disease must also be prevalent.

Symptoms.—Lassitude, dull and heavy manner, appetite lessened, shiverings, dryness of the nose, and frequent sneezing are the earliest symptoms observed. Redness and swelling of the nasal mucous membrane soon supervene, and an increase of the secretion begins. The eyes become reddened and watery, the nose hot. Fever is often associated with these symptoms, and the pulse runs above the normal. In simple uncomplicated catarrh, these are the prominent manifestations. When cough, changed respiration, and other symptoms appear, the affection has extended, its identity is lost and immerged into other diseases.

Treatment. — A simple cold should never be neglected. Uncomplicated, it is self-limited and naturally tends to recovery. The dog should be placed in warm comfortable quarters, and a dose of castor oil or syrup of buckthorn be given. If much fever is observed obtain the following mixture. —

> R   Tr. Aconiti Rad.   gtts. xij
>   Spts. Ætheris Nit.   ℥ ss
>   Liq. Ammon. Acet.   ℥ iss
>
> Ft. Mist.   Sig. Dose two teaspoonfuls every three hours in water.

At night five grains of Dover's powder should be administered. If the nasal secretion becomes thick, viscid, and not easily expelled, the fluid extract of hamamelis diluted with four parts of water may be injected into the nose.

Milk and broths for the first day or two will be all and possibly more than the dog will care to eat. Should his appetite not return soon however, scraped raw beef should be given in small quantities, three or four times daily with two grains of quinine in pill form.

Nasal Polypus.

Occasionally mucous polypi of the nasal cavity are seen in dogs. These tumors retain the elements of the mucous membranes from which they arise. Certain catarrhal symptoms are associated with their formation, but their presence can rarely be positively determined until they can be plainly seen. They are a soft jelly like mass which in time fill up the entire nasal passage, closing the opening. Polypi arise and run their course without pain.

Treatment. — One author says, " the treatment is simple, and consists in the removal of the polypus, by tying a strong silk thread or piece of fine silver wire round the neck. This must be tightened daily for a few days till the neck is cut through and the tumor drops off." Very excellent advice and apparently a ready solution of the whole matter. But let the operation be undertaken, and the difficulty between theory and practice will be at once appreciable.

The writer advises when a polypus exists, that its removal be entrusted to a physician, who will in many cases need all his skill and experience in performing the operation.

OZÆNA.

It is stated by some authors "that ozæna is an affection rarely seen in the dog." While the severest forms of the disease are certainly seldom met with, the milder types it is believed are more common.

Causation. —A low state of the system, impoverished by neglect or disease invites the affection. Under such conditions, the cause is usually found in successive attacks of catarrh, and sometimes in distemper. Diseases of bone, ulcerations, foreign bodies, and abnormal growths, like polypus, within the nose are commonly associated with ozæna.

Symptoms. —The secretion in this disease varies much in quantity and appearance. The amount in some instances is profuse, in others scanty. The discharge is of a purulent character which decomposes, and a peculiar stench is produced giving rise to the name of the disease. The discharge has a tendency to crust formations deep within the nose, as well as about the outer openings, which are usually found coated with a greenish deposit. Constant sneezing is a marked symptom and the breathing through the nose is considerably obstructed.

Prognosis. —If the disease is of recent origin it can be cured but the gain will be slow. Much will depend upon the general state of the system and its reparative powers.

Treatment. —Efforts must be directed to restore the mucous membrane to its normal condition. Local treatment is the first essential. The nose should be injected every day with the following mixture. —

℞	Acidi Carbolici	gr. xx
	Acidi Tannici	gr. xx
	Glycerinæ	℥i
	Aquæ	℥iij
	Ft. Mist. Sig. Inject.	

Quinine, iron, or cod liver oil should be given internally, as the case demands, and a generous diet allowed.

INFLUENZA.

Influenza is an epidemic disease, which is characterized by catarrh of the respiratory, and often also of the digestive organs. Its attack is sudden, great and rapid weakness is developed, and fever of greater or less intensity is associated. The disease may occur among dogs, uninfluenced by the age of the animal or the season of the year.

Causation. — Of the exciting causes of influenza little is known. It has not been established that there is any connection with atmospheric conditions. An essential difference is noted between this and the non-epidemic catarrh, since the latter affection is traceable to "catching cold," and follows exposure. Influenza on the other hand, prevails when catarrhal affections are rare, and it may appear in any season of the year, hot and dry, or cold and damp. Some authors assume it to be contagious, others refute the theory, and the question has remained a mooted one.

In 1872 an influenza prevailed widely among horses, and in New York alone attacked about sixteen thousand; no special causes for it were ascertained.

Symptoms. — The frequent location of the disease is upon the mucous membrane of the respiratory organs, it being more rarely affected than that of the stomach and intestines.

The onset of the disease is generally sudden, marked by lassitude and shiverings, indicative of a chill. In severe cases fever soon manifests itself assuming a remittent type. The pulse varies in character being full and frequently irregular. The affection of the mucous membrane of the respiratory organs is indicated by symptoms essentially of a catarrhal nature. Most frequently is observed at first sneezing, the eyes reddened and watery, the nasal membrane swollen and reddened.

In some cases the throat becomes sore, and the catarrhal inflammation extends to the larynx and bronchi, the invasion of these parts being evinced by hoarseness and a tendency to cough. Not unfrequently in the very beginning, a cough sets in dry and disturbing, occurring in paroxysms frequently inducing vomiting. The expectoration is usually scanty but sometimes becomes more abundant later in the disease. While the physical examination of the chest is negative, the breathing frequently becomes hurried and at times the restlessness of the animal will indicate pain, probably of a pleuritic character. The digestive organs are very often affected. Loss of appetite, a coated tongue, thirst, and not infrequently vomiting and diarrhœa accompany the disease. At times however, constipation exists.

Prostration is often very marked, the animal is dull, heavy, and sleepy. It will be frequently noted that in walking about he will act stiff and lame and sometimes his movements will appear painful.

The expression of the eyes is changed in a marked degree, wearing a depressed anxious look. Late in the disease there is a tendency to dropsy of the limbs, and sometimes an effusion into the chest.

Diagnosis. —To distinguish between other affections resembling influenza should not be difficult; the existence of an epidemic, other dogs being attacked at the same time; the great prostration from the first; the fever of a remittent type; the peculiar cough and absence of signs indicative of disease of the lungs. A consideration of these facts should make a differential and positive diagnosis easy.

Prognosis. —If uncomplicated, the disease commonly ends in recovery. In old dogs and in very young puppies it runs a more severe course. The existence of other diseases, and a weakened and impoverished system would render a prognosis more unfavorable.

Treatment. —The animal should be placed in a room of even temperature and well ventilated, but free from draughts. In mild cases the medicinal treatment need not be especially active. The congestion of the head can be relieved by gentle laxatives, but purgatives are contra-indicated. If the fever runs very high, quinine in three grain doses should be given every four or five hours. When the temperature has become normal the quinine may be continued in two grain doses, twice or three times daily for a time. If the larynyeal inflammation is severe, hot applications should be made to the neck, or the same rubbed well with the ammonia and camphor liniments in equal parts, and afterwards swathed in cotton or flannel.

In extremely severe cases where the breathing becomes difficult, the animal may be made to inhale steam; the water from which it is generated should be tinctured with carbolic acid, two drachms to the pint. An admirable method of keeping the air the animal breathes constantly disinfected, is to use some form of lime and carbolic acid mixed together.

An important consideration in treatment is to guard against possible failure of the vital powers, especially in old dogs and others weakened by diseases. Quinine or iron should be persisted in, and the following mixture be given. —

> R   Ammoniæ Mur.   ℥ij
>   Ext. Glycyrrhizæ Pulv.   ℥ij
>   Aquæ   ℥iij

Ft. Mist. Sig. Dose one teaspoonful in water every three hours.

The diet should be easily digestible; milk, broths, and if needed, eggs added to the former.

CHAPTER V.

AFFECTIONS

OF

THE MOUTH AND TONGUE.

THE ERUPTION OF THE TEETH.

The first or milk teeth of a puppy begin to appear soon after birth, and within a month the first process of dentition is completed. The permanent about the fourth month, begin to replace the temporary teeth, and are usually through the gums before the eighth month. They are forty two in number and the following is the recognized formula.—

$$\left.\begin{array}{l}\text{Upper jaw—Incisors 6 ; Canines 1—1 ; Molars } 6—6=20 \\ \text{Lower jaw— \ \ „ \ \ 6; \ \ „ \ \ 1—1; \ \ „ \ \ 7—7=22}\end{array}\right\} = 42$$

Occasionally supernumerary teeth appear and are irregularly placed.

It has been supposed that by the teeth of the dog his age could be determined; a mistake in many instances as is very evident. Disease will cause them to decay, certain medicines promote their early destruction, the too frequent feeding of bones wear them away, and a young dog may have poor teeth, and those of an old one be well preserved.

It will be observed by the arrangement of the teeth that their purpose is to bite, tear, and crush. Nature makes no provision for grinding or mastication. In this the construction is identical with that of all carnivorous or flesh eating animals. The food is bolted, and passes into the stomach without meeting with either a mechanical change from the teeth, or a chemical influence from salivary secretions of the mouth.

TREATMENT OF THE TEETH.

Generally but little trouble is experienced by puppies in shedding their temporary teeth. It would be wise, however, to examine their mouths occasionally, and if any are loosened remove them, which can easily be done by pressing on the side of the tooth.

Fits are sometimes attributable to teething. When they occur, if the gums seem swollen, it would be wise to remove all doubts by lancing.

If a tooth becomes decayed and evidently painful, the dog should not be tortured by the unskilful use of primitive "tools," but rather a trifling expense be incurred and the operation left to a professional dentist.

If the diet of dogs is properly chosen and bones are judiciously fed, the animals will clean their own teeth and obviate the need of their owner's intervention.

Tartar is an earthy matter deposited on the teeth from the saliva. If suffered to accumulate, it causes inflammation and absorption of the gums and gradual loosening of the teeth. When any quantity has accumulated it should be removed with the blade of a pen knife, and for a time the teeth cleaned daily with a tooth brush and charcoal; then bones allowed once or twice a week, the gnawing of which will insure protection from the deposit.

SOFTENING OF THE GUMS.

This condition may be the consequence of an accumulation of tartar, or of a congested state of the liver and bowels.

The gums are swollen, spongy and tender, and bleed at the slightest touch. In such cases the breath is usually very offensive, and other evidences of bilious and digestive disturbances are apparent.

The affection is more commonly met with in house pets, over indulged and deprived of sufficient exercise. The dispositions in such cases become materially changed, the animals growing obstinate and sulky. A condition absurdedly attributed to the giving of meat, whereas, it is due to confinement and an absence of common sense in feeding. Such dogs it will be observed, have very little appetite, preferring dainties and condiments rather than wholesome food.

Treatment. —Dietetic influences are the first to be observed. In a recent publication one author states: "it used to be the custom of the older practitioners to commence the treatment of such cases by two or three days' starvation. This treatment is terribly cruel, and any one who should prescribe such should be punished by law." The writer believes the "old practitioners" did not go so very far astray in their treatment. Two days denial of food, in such cases, is far from " cruel" but rather merciful. Few indulgent owners can be convinced of this however, and the extreme must be modified.

The treatment may be wisely commenced with a dose of castor oil or syrup of buckthorn or both. No more than two meals daily at the utmost should be allowed, and at each a little milk be given at first. This may be followed by broths, and raw meat occasionally.

Exercise must be strictly enforced, the animal being walked a mile in the beginning, then two miles a day, later run behind a carriage. As he becomes stronger and more active the diet may be more generous, meat in some form entering largely into it.

The teeth should be looked to, tartar removed, and the gums rubbed with a little tincture of myrrh. Patience and common sense are certain to overcome the affection unless the dog be too old.

CANCRUM ORIS.

Canker of the mouth is an individual affection induced directly by a decayed tooth, or a diseased state of the jaw bone.

Symptoms. —Attention is usually first attracted by a swelling on the jaw over or beneath the part diseased. This swelling finally bursts, discharging matter and blood of a very offensive odor. The parts affected are very painful, and usually much difficulty will be experienced before a thorough examination is made.

After the abscess has burst, a fistulous opening remains that will never permanently close until the diseased bone or tooth is removed. A discharge will persist having the same offensive character. Eventually, a fetid ulcer forms on the gums and cheek of a gangrenous character, bleeding easily at the slightest touch. The pain and difficulty in eating tend to exhaust the animal which becomes weakened and emaciated.

Treatment. —Whatever the cause it must first be removed. If a decayed tooth and it is extracted, the trouble will probably disappear. If the bone of the jaw be diseased, then the dog should be etherized, the gum laid open and the affected portion removed. The bleeding growths and ulcerations may then be well burned with the nitrate of silver and afterwards be treated with the tincture of myrrh. The general health must not be overlooked. The diet should be generous; milk, broths, raw eggs, meat, and if need be, tonics must be administered; cod liver oil in tablespoonful doses, and if digestive disorders are present, the elixir bark pepsin and bismuth, one teaspoonful three times daily.

SALIVATION.

Salivation or an increased flow of saliva, may be caused by a too liberal use of mercury, by a sudden check to the cutaneous secretion, by cold and damp, or by any influence that suddenly lowers the system. The disease may also arise from decayed teeth. It is commonly associated with cancrum oris and may be induced by some irritation of the salivary glands alone. Probably the more common cause of salivation is the improper use of mercury, either taken internally or used too freely externally. In prescribing the drug to dogs, it must be remembered that natures vary. Some are far more easily affected by its action than others, and a susceptibility of salivation and tendency to certain diseases often go together.

Symptoms.—Swelling and inflammation of the salivary glands, tongue and throat, with a flow of peculiarly fœtid saliva. Shallow ulcerations of the surface of the gums and of the inside of the cheeks. The gums are red, swollen and spongy. The teeth loosened and discolored. In the severest cases, the mucous membrane of the stomach and intestines takes on much the same ulcerative action observed in the mouth.

The failure of strength and loss of flesh depend upon the exciting agency of the affection as well as the severity of the symptoms.

Treatment.—Decayed teeth should be extracted; the bowels kept open and a generous diet be allowed. For a local application the tincture of myrrh, a saturated solution of chlorate of potassa, or brandy and water is sufficient.

If the cause of the salivation is mercury, the iodide of potassium in three grain doses, three times daily, should be administered. If no cause for the affection other than debility is apparent, tonics of iron, quinine, or cod liver oil should be persisted in, with the local applications.

BLAIN.

Blain, manifested by an eruption of blisters on the tongue, is comparatively rare among dogs.

The cause is generally attributable to improper feeding, neglect, and exposure, and other influences impoverishing and debilitating the general system.

Symptoms.—The blisters occur on the sides and the under surface of the tongue, continue as such for a short time, then burst, leaving ulcers to occupy their place. These ulcerations throw off a purulent, bloody discharge having a peculiar fœtid odor. Some degree of salivation is associated, and the tongue is swollen and inflamed. Often the tissues forming the floor of the mouth are involved in the inflammation, and abscesses result. In many cases the glands of the neck become enlarged and painful.

Treatment. — Chlorate potassa is a remedy of the greatest efficacy in this disease. As debility, if not the cause, will very likely result as a complication, the early administration of tonics is advised. Chlorate potassa and iron are combined as follows. —

R   Potass. Chloratis   ℥ ij

   Tinct. Ferri. Chloridi   ℥ i

   Syr. Aurantii   ℥ i

   Aquæ   ℥ iij

Ft. Mist. Sig. Dose one teaspoonful every one or two hours.

Equal parts of the flowers of sulphur and white sugar should be mixed in fine powder, and after sponging out the mouth, this powder should be dusted over the ulcerations three or four times a day. Abscesses if occurring should be lanced when ready.

Nourishing and concentrated food of a liquid character is demanded.

INFLAMMATION OF THE TONGUE.

This affection is known by great swelling of the tongue, tenderness and difficulty in swallowing. It generally accompanies severe salivation, but it may occur in an acute form independent of this cause.

In convulsions the tongue is often bitten and severe inflammation follows. The disease may run a rapid course and abate without the formation of "matter;" on the other hand, abscesses may form under the tongue, sufficient in size to threaten suffocation.

Treatment. — The bowels should be freely acted upon by aperient medicine. Applications of ice or ice water made continually for hours promise relief. Chlorate of potassa or borax should be added to the water making a strong solution. If suffocation is threatened, a surgeon should be called to make an incision in the seam running along the middle of the tongue on the under surface.

In event abscesses form; they should be opened when the proper stage is reached. A skilled hand alone should undertake the operation, or severe hemorrhage may result.

The food must necessarily be liquid and concentrated. A cup of milk into which has been beaten a raw egg may be given every three or four hours as the need of support is apparent.

PHARYNGITIS.

The pharynx, a part of the canal through which the food passes, is placed behind the mouth, nose, and larynx.

Inflammation of the pharynx, or pharyngitis, is an affection very rarely occurring alone, but is usually associated with some other disease. When it does appear as an individual affection it is commonly caused by the lodgement of foreign bodies, such as sharp pieces of bone. When any substance becomes fixed in the pharynx, about the larynx, or in the top of the passage into the stomach, it may produce a sense of choking and fits of suffocative coughing, or symptoms less severe. A peculiar throat cough with frequent hawking is often the prominent manifestation at first.

Later on in the affection, the mucous membrane may become swollen, congested, and swallowing difficult, the throat being sensitive to external pressure, and coughing excited by it as in laryngitis.

Sometimes, but very rarely, abscesses will form in the back part of the pharynx. Their presence may be suspected if pharyngitis has existed for several days, and the animal carries his head extended, fixed and immovable on his neck.

Treatment. — Foreign bodies should be sought by exploring the throat with the finger. If removed at once no further treatment is necessary.

If repeated efforts fail to dislodge the substance, an emetic of ipecac or sulphate of zinc, dose of either fifteen grains, should be given to induce vomiting. The treatment of the pharyneal inflammation should be the same as recommended in acute laryngitis.

CHAPTER VI.

DISEASES AFFECTING

THE

DIGESTIVE SYSTEM.

INDIGESTION.

A large proportion of the disturbances of the stomach and intestines relates to digestion. In every attack of indigestion the processes involved in the changes which the food undergoes before absorption, are not only retarded, but also accompanied by numerous local and general difficulties. Indigestion, or as termed by some authors dyspepsia, was formerly supposed to be a purely functional affection, existing without any morbid changes in the gastric mucous membrane. This theory is not generally accepted at the present time.

The name indigestion as herein used does not apply to disorders incident to fevers, inflammatory affections, or structural diseases of the digestive organs.

Causation.—The influences which induce this disease may act directly or indirectly. Food if irregularly given in excess or of an unwholesome quality, too long fasts, and weakness of the digestive organs following inflammations are among the direct causes.

The indirect agencies involved are far more numerous. Want of sufficient exercise, over exertion, poverty of the blood, and disarrangements of the system are among the more prominent causes.

Symptoms.—Probably the most constant sign of indigestion is impairment of the appetite, which is at least capricious if not entirely absent.

The food taken into the stomach digests very slowly, and gases are generated, producing distension or bloating. Vomiting occurs; masses of mucus and food are rejected in a more or less altered condition.

The matters vomited are occasionally mixed with bile, which is an unfavorable complication, as its presence in the stomach additionally disturbs digestion. The appearance of streaks of blood at times has no significance whatever. In an attack of indigestion, changes occur in the digestive tract both above and below the stomach. The former are shown by thirst, a coated tongue, hot nose, increased salivary secretion, and a foul breath. Below the stomach, associated disorders are manifested by constipation or diarrhoea, abdominal distension, and colic.

The animal is often dull, listless, and depressed in his manner, and in severe cases fever in a certain degree, is present.

Treatment.—In some attacks of indigestion a regulated diet suffices to control the disturbance. When the disease is of a more severe nature, the stomach must be favored as much as possible, and only the simplest and most easily digested articles of food given.

In commencing treatment it is desirable to remove any portions of food retained in the stomach and fermented, the presence of which keeps up the irritation of the gastric mucous membrane. This can be done by emetics or laxatives. If persistent vomiting has existed, naturally emetics are contra-indicated; the same with laxatives if diarrhoea is present, unless the discharges are very offensive in character, in which instance a cathartic need not be withheld, but would rather be indicated to remove the exciting cause evidently still acting. The wine of ipecac is recommended as an emetic, two teaspoonfuls of which should be given as the first dose, followed by one teaspoonful every ten minutes until vomiting occurs.

From its peculiar action in neutralizing the abnormal quantity of acid in the stomach, calcined magnesia is the better laxative to employ, and one teaspoonful should be given in milk.

After the stomach has been emptied of its irritating contents, the greatest care should be exercised in feeding, and only the simplest and most easily digested articles of food chosen. In the severest cases, nourishment should be given every two or three hours. No better diet can be advised than scraped raw beef, raw eggs, and a little gelatine mixed together, and given in tablespoonful quantites. As the stomach grows stronger the diet may be varied by giving milk and lime water, bread and milk, beef tea, meat stews well cooked, flavored with a little onion and thickened with rice, corn, or oatmeal, the animal allowed to eat of the meat chopped fine. It must be remembered that the liability to vomit is increased in proportion to the amount of food taken into the stomach.

In the use of medicines in indigestion, the peculiarities of the animal must be understood, and experimental treatment as in feeding may be necessary.

The writer submits a variety of drugs and combinations applicable to the different indications and phases of the disease; if the use of one proves unsuccessful another should be substituted.

In cases of vomiting, from five to ten grains of the subnitrate of bismuth should be placed on the tongue, and washed down with a tablespoonful of milk before nourishment is taken.

With the food should be mingled either of the following agents — French pepsin three to five grains, the saccharated pepsin five grains, one peptonic pill, or five grains of lactopeptine. When pure pepsin can be secured it is undoubtedly the most admirable agent which can be used to assist digestion.

It must be remembered however, that during the past few years the quality has sadly degenerated, and rarely can a reliable preparation be obtained. The saccharated is many times inert, and even the French pepsin the most expensive form, will often be found inactive. The pill peptonic and lactopeptine are more recent preparations, and will probably for a time at least prove reliable.

As the disease is in a measure controlled, and the need of such frequent feeding and dosing is no longer manifested, the number of meals can be lessened to three, continued for a time, and then restricted to two. The tonics which may be advantageously used are many; among them the elixir bark pepsin and bismuth, the elixir calisaya iron and bismuth, of either a teaspoonful before eating. The chlorate potassa is sometimes efficaciously employed and combined as follows.—

R	Elixir Calisayæ Ferri et Bismuth.	℥ ij
	Potass. Chloratis	℥ ij
	Aquæ	ad. ℥ iv

Ft. Mist. Sig. Dose one teaspoonful before eating.

Strychnine is a useful agent and may be given with pepsin and bismuth as follows.—

R	Pepsini Sacch.	℥ iss
	Bismuth. Subnitratis	℥ iss
	Strychniæ	gr. ss

Ft. Pil. No. xxx. Sig. Dose one, three times daily before eating.

Where constipation exists with gaseous distension, the following is a useful preparation.—

R	Sodæ Bicarb.	
	Pulv. Rhei	
	Sacch. Pepsini	
	Pulv. Zingiberis	
	Bismuth. Subnitratis	
	Pulv. Cubebæ	aa ℥ iij

Ft. Pulv. Sig. Dose from one third to one half a teaspoonful after eating, in milk.

At times in cases of indigestion, it will be observed that the discharges from the bowels are clay-colored, an evidence of liver derangements or closure of the passage from the gall-bladder to the intestines. No especial medication for this condition is demanded unless constipation co-exists, in which case it will be well to employ a tonic, with which a laxative having a direct action on the liver is combined, as in the following. —

    ℞ Quiniæ Bisulph.   ʒss
     Ferri Sulph. Exsic. ʒss
     Acid Arseniosi   gr.ss
     Ext. Taraxaci   ʒi

Ft. Pil. No. xxx. Sig. Dose one, two or three times daily before feeding.

The diarrhoea which sometimes accompanies indigestion calls for no especial treatment, unless evident pain attends the discharges, in which case a teaspoonful of paregoric given occasionally as demanded, will prove sufficient.

Recovery from severe cases of indigestion may be slow and tedious, but the result of patience and judicious treatment in the end, well repays the effort and waiting.

ACUTE GASTRITIS.

It should be understood that the term acute gastritis is used to denote an inflammation of the stomach, of such intensity as to greatly endanger life; a disease fortunately of very rare occurrence. It is very doubtful if this affection ever originates from the internal and unappreciable causes which give rise to the so-called spontaneous inflammations in other situations. The cause of the disease may not in every instance be obvious; it can however, with almost certainty be considered traumatically produced by the action of caustic or irritant poisons.

Symptoms.—Pain and vomiting are prominent symptoms in acute gastritis, the stomach being often intolerant of water or the blandest liquids. The matter vomited is a watery mucus of a greenish color, not unfrequently mixed with blood dark and clotted in appearance. The acts of vomiting are evidently painful. The desire for cold drinks is excessive, notwithstanding the distressing acts of vomiting it provokes. Pressure over the stomach reveals tenderness.

The position assumed by the animal is characteristic; it stretches itself out on its belly; the pulse is up and fever is more or less intense; the expression is anxious and haggard; the bowels might be constipated but are more commonly relaxed and diarrhoea is present. The nose is dry, and if the inflammation is severe the respiration is hurried.

If death does not speedily occur and the disease runs on to a fatal termination, dark clotted blood will be vomited. Everything taken into the stomach will be expelled almost without effort. Prostration becomes evident; the pulse grows rapid and thready; the limbs become cold, and the skin clammy. Convulsions sometimes occur and finally the poor victim dies from exhaustion. The end is usually very rapid in fatal cases.

Diagnosis.—There is little danger of acute gastritis being mistaken for any other affection. With a view to appropriate treatment, it is important to determine from the symptoms whether or not the attack is positively attributable to poison. If the animal is taken suddenly ill and his previous movements and whereabouts can be learned, it will be of some assistance. The mouth and throat should show some trace of the poison if of a corrosive character, of which an early vomiting of blood in quantities is markedly indicative.

Prognosis.—Acute gastritis is undoubtedly a disease of great gravity. It may prove fatal within a few hours, the animal dying from shock; or the end may come slowly, the result of exhaustion. The amount of local injury will in a great degree measure the danger.

Treatment.—It is to be inferred that in cases of poisoning, the proper antidotes have been administered, and the contents of the stomach evacuated. The next step is to secure as much as possible rest for the inflamed organ. For this purpose morphine is to be relied upon, one grain of which should be divided into twelve powders. One of these may be dropped on the tongue and washed down with a teaspoonful of milk, every two or three hours at least; oftener if the pain seems severe and not inclined to subside.

If the vomiting is persistent and the morphine given in this manner evidently has little or no effect, it should be administered subcutaneously. When this method cannot be employed for want of a proper instrument, an anal injection should be resorted to, using a glass syringe, and throwing into the bowel two morphine powders or thirty drops of laudanum in a little milk. Pressure must be maintained at the anus for some time lest the injection be returned. This treatment can be repeated every two hours if the indications warrant.

If the symptoms denote a tendency to exhaustion, supportive measures must be employed. The attempt should first be made to feed by the mouth with small quantities of beef tea, raw eggs, and gelatine mixed. Milk with the addition of lime water, arrow root, and flaxseed tea are sometimes well borne.

It must be remembered that all food should be iced and given in quantities of not over a tablespoonful, less if not well retained, and repeated every ten or twenty minutes if necessary. If this treatment does not sustain the animal, then eggs, beef extracts, etc., with brandy should be added to the injection of laudanum.

After the inflammation has disappeared, the stomach will be left in a very debilitated condition, and remedies of a tonic character be indicated. The elixir calisaya iron and bismuth, in teaspoonful doses three or four times daily is advised.

CHRONIC GASTRITIS.

Chronic inflammation of the stomach is not an unfrequent disease among dogs. It is due to a congestion, and a low type of inflammation of the mucous membrane, and is usually combined with disorders of other organs, more especially the liver.

Causation. —Chronic gastritis may follow the acute disease. It may also proceed from excesses in eating, or long fasting.

Andral found the stomach inflamed with ulcerations in dogs, destroyed by starvation.

Symptoms. —The symptoms of indigestion are present in this disease. Tenderness over the region of the stomach very often exists. Thirst is constant and vomiting distressing. Pain in many cases is prominent.

Diagnosis. —To distinguish between indigestion and the disease under consideration is not always easy. In the former the appetite is less impaired, the weight and nutrition better sustained, vomiting, fever, thirst, and pain are less prominent symptoms than in the latter disease.

Prognosis. —Recovery is often slow and tedious. Patience, careful nursing, and judicious feeding will aid much the action of medicine and promote the chances of recovery.

Treatment. —Much the same treatment should be employed as in indigestion, especially in the matter of diet. Purgatives are to be avoided. If the bowels become constipated an injection of soap-suds will be sufficient. External applications are indicated, and small fly-blisters will be found efficacious. Internally the most valuable remedy is nitrate of silver, and may be combined with opium as follows. —

℞ Argenti Nitratis   gr. iv
    Pulveris Opii   gr. vi

Ft. Pil. No. **xxiv.** Sig. Dose one three times daily.

Bismuth is a most admirable agent in this disease, and may be added to the nitrate of silver as in the following. —

℞ Extract. Hyoscyami   gr. xv
   Argenti Nitratis   gr. iv
   Bismuth. Subnitratis ʒi

Ft. Pil. No. **xxiv.** Sig. Dose one three times daily.

VOMITING.

As it is a dog's license to vomit easily and almost at will, an attack may mean much, and it may be of the least importance. Often a hearty meal will be bolted to be immediately expelled, and again more leisurely eaten, a mere case of overloading the stomach.

Vomiting is frequently associated with pulmonary disorders, caused by thick mucus lodging in the throat. It is a very common symptom in diseases of the stomach and intestines, oftener the former.

Matter vomited may be undigested food or mucus alone, the latter a more urgent symptom. Blood may be vomited if there is much straining, and is unimportant if the quantity is small. When it is fresh, bright red in color, and ejected in quantity, the cause may be a sharp bone lodged in the passage to the stomach, and the lining membrane lacerated.

Gastric ulcer may be suspected when a severe hemorrhage occurs, and it is more probable if a previous attack has been experienced. If a dog vomits often within a short time, bile will frequently appear mingled with the matter raised; its presence then is purely mechanical, the continued retching drawing it up into the stomach, instead of permitting it to flow down the intestines. Vomiting is symptomatic of a variety of affections exclusive of inflammation or structural diseases of the stomach.

Treatment. — The cause should be studied and if possible removed. If occasioned by indigestion, worms etc., appropriate remedies are directed in the treatment of those diseases. In functional disturbances of the stomach, vomiting may be considered as an individual affection and as such demands especial recognition. The following mixture is of value. —

R   Bismuth. Subnitratis   ʒiij
    Acidi Hydrocyanici dil.   ʒi
    Aquæ   ʒv

Ft. Mist. Sig. Shake well. Dose one teaspoonful three or four times daily.

Carbolic acid is valuable as a remedy and can be given as follows. —

R   Acidi Carbolici   gr. vi
    Glycerinæ   ʒij
    Spts. Lavand. Comp.   ʒi
    Mucil. Acaciæ   ʒiss

Ft. Mist. Sig. Shake well. One teaspoonful every two hours while vomiting.

At times the use of opiates are demanded, given as follows. —

R   Morphiæ Sulphatis   gr. i
    Cerii Oxalati   gr. xx
    Sacch. Albæ   ʒss

Ft. Chart. No. xij. Sig. One in a little milk three or four times daily.

COLIC.

Colic is a pain that originates in the walls of the intestines. It is properly but a symptom of the many abdominal diseases; its importance and frequent appearance among dogs however warrants its classification as a distinct disease. Among animals an attack of colic has especial significance from the fact that inflammation is prone to speedily follow it.

Puppies are especially liable to the affection near the period of weaning, and petted house dogs and pampered pets seem oftener attacked.

Causation. — The influences which give rise to the affection are many. Among them indigestion, gaseous distension, constipation, worms, the passage of calculi, and intestinal obstruction or stoppage. Colic pains occur with diarrhoea that has originated in consequence of irritating substances being taken into the stomach. Cold and exposure may induce the affection. The introduction of lead into the system in poisonous amounts will also develop a special form of the disease.

Symptoms. — Restlessness and occasional moanings, or sharp cries of pain are the first to attract attention. It will then be observed that the back of the animal is arched, the abdomen hard, and the walls retracted. Firm pressure will often cause shrinking and evident pain.

The sufferer lies down only to be up and walking about again, vainly seeking relief by constant change of position. The manner plainly indicates the suffering; the expression is piteous and appealing. If the attack is severe moaning is constant, varied only by sharp, ear-piercing cries.

The animal if a petted house-dog, will seek his master or mistress as though begging for relief; unless it comes quickly his manner changes, he seems to shun all interference and endeavors to conceal himself in retired and unfrequented places.

At first no constitutional symptoms are associated; the pulse, nose, and skin remain normal. If the attack is prolonged evidences of weakness and fever are manifested. Vomiting and ineffectual attempts to vomit are common symptoms. The appetite usually entirely disappears.

Diagnosis. — The suddenness of the attack, and the absence of all signs of inflammation render a diagnosis easy.

Treatment — Although writers agree that the first step in treatment is to administer a cathartic, it seems ill advised, and a better practice is to at once attack the pain, the more urgent symptom. Opium in some form is the only agent upon which we can depend in the severer cases. Paregoric and laudanum are both preparations of opium, and one or the other is to be found in nearly every house. If the former is used, one teaspoonful should be given from one to three hours until the pain is subdued. Of laudanum the dose is fifteen drops repeated in the same way.

Hot applications may be made to the bowels and will prove grateful to the sufferer. If vomiting occurs at intervals and the medicine is not retained, twice the quantity given by the mouth should be injected into the rectum and held there for a time by pressure over the anus. If the restlessness of the animal makes this effort ineffectual, a subcutaneous injection of one eighth of a grain of morphine should be administered.

If a chemist is within easy distance instead of paregoric or laudanum, have the following prepared.—

   ℞ Chloroformi   ℥ss
     Elixir Mc. Munn's ℥iss
     Ext. Pruni Virg. Fld. ad. ℥iv

Ft. Mist. Sig. Dose one teaspoonful in a little water every half hour until the pain ceases.

The causative treatment, having for its object the removal of the different causes, should be commenced as speedily as possible. A free injection of warm soap-suds should be given to invite a movement of the bowels by exciting intestinal activity.

As soon as possible after the pain has been subdued a tablespoonful of castor oil should be administered and the discharges watched for, to determine the cause occasioning the attack. If attributable to coarse indigestible nutriments, a diarrhoea will probably follow, yielding readily to proper dietetic restrictions.

One fact will be observed and surprise will be felt at the amount of opiates some dogs will bear without visible effect.

Following an attack, the diet should be restricted and the same treatment instituted as advised in Indigestion.

DIARRHOEA.

The term diarrhoea is used to denote too frequent operations of the bowels, the discharges being abnormally changed in character. It may occur as an independent affection, or as a symptom incident to other diseases.

It is often associated with inflammation of the large and small intestines, and with disorders where changes in structure exist. It will also appear in functional derangements, such as indigestion and colic.

Causation.—The exciting causes of diarrhoea are extremely numerous. When the affection is purely functional, the mucous membrane being in a normal state, among the causes are included indigestible substances, decayed food, foreign bodies, retained excretions, poisons, and medicines. Worms lodged in the lower bowel may accumulate and diarrhoea follow.

Another common cause is the influence of cold, intestinal irritation resulting from the cooled blood of the surface being driven inward. Traumatic injuries such as a kick or a blow in the abdomen, may induce the disorder. Diarrhoea is often produced in puppies by feeding milk when they are unaccustomed to it. On the other hand, after giving it for a time it may be found to produce constipation. Radical changes in feeding, from coarse food of difficult digestion to a more nutritious and concentrated diet, will loosen the bowels.

In diarrhoea associated with other diseases, the intestinal mucous membrane becomes inflamed from causes attributable to the associated disorder, as in affections of the liver. According as the exciting cause of diarrhoea is of a temporary, frequently recurring, or permanent character, the affection runs an acute or a chronic course.

Symptoms.—The frequency of the discharges will in a measure indicate the severity of the disease. More important than the number is the character of the stools. Blood is not an uncommon mixture. It may be due to a congested state of the membrane lining the lower bowel, to piles, to a sharp bone lodged in the rectum, or possibly worms. The presence of mucus in small amounts has no significance. In large quantities an inflammation of the rectum is to be inferred.

In simple uncomplicated diarrhoea, unless of long continuance the general health suffers but little.

Treatment.—Diarrhoea in many instances is nature's effort to throw off the cause of irritation, and regulating the diet will usually suffice to overcome the trouble. If the attack is severe, nature should be imitated and a dose of castor oil given, to effectually remove the contents of the intestines, preventing their continued passage over the irritated surfaces.

If pain exists, fifteen drops of laudanum should be added to the laxative.

The diet then should be regulated, being simple and easily digestible ; milk with lime water in parts three to one, or milk porridge made with flour baked until browned. If this is refused, beef broths thickened with raw eggs will be more inviting. The elixir of bark pepsin and bismuth should be given in teaspoonful doses before food is taken. This remedy and the dietetic precautions will in many instances suffice ; even in serious cases the treatment should be given a fair trial before resorting to stronger medication.

If the disease persists and more urgent methods are indicated, opium should be employed. Dover's powder in five grain doses repeated every five or six hours is admirable on account of its combination. If that proves insufficient and pain is still a prominent symptom, the following may be given.—

R	Tinct. Kino	℥ ss
	Tinct. Opii	ʒ ij
	Syr. Simp.	℥ ss
	Chloroformi	ʒ i
	Aquæ Menth. Pip.	℥ ijss

Ft. Mist. Sig. Dose one teaspoonful from four to six hours as indicated.

It is quite often necessary to repeat the castor oil ; the character of the discharges will direct the need. If they are of a very offensive odor then a dose should be given every two or three days.

Diarrhoea in young puppies very often occurs soon after the period of weaning ; changing the diet will often suffice with them. If feeding milk freely, give less of it and more beef tea, and vice versa; if this method is ineffective, chalk mixture in teaspoonful doses should be given often, and five or ten drops of paregoric be added if pain evidently exists.

DYSENTERY.

The term dysentery is used to denote an inflammation of the large intestine accompanied by mucous and bloody discharges from the bowels.

In mild cases the inflammation is not severe and may be confined to the rectum. In severe attacks of the disease, the inflammation is not only intense but extensive, involving the greater part of the large intestines, the mucous membrane of which becomes reddened, swollen, and ulcerated. The ulcers are greater or less in number, some being small and others of considerable size.

Causation. — Excesses in eating, food decomposed or improperly cooked, foul drinking water, exposure to cold and rain storms are among the generally recognized causes. In many cases it is not easy to trace the origin of the disease under consideration to any obvious agency.

Symptoms. — The affection is generally preceded by a diarrhoea of a variable duration, with which some pain has been associated. The appetite is lessened and the manner dull and listless.

The development of the disease is denoted by characteristic discharges consisting of mucus with which more or less blood is commingled. The effort to move the bowels is frequently made and the quantity passed at each act is generally small. Slight evacuations may take place every hour and even much oftener.

The quantity of mucus expelled is in some instances quite abundant and appears in a jelly-like mass, for which the popular term applied is slime. Sometimes fecal matter is mingled with the discharges but they are more commonly dysenteric in character, consisting of mucus and blood. At times the evacuations present a greenish color. The amount of mucus and blood voided constitutes measurably a standard for judging the extent of the intestinal surface affected.

Pus sometimes appears in the discharges in the acute form of the disease, but more commonly it is observed in the chronic stage. The inflammation of the rectum occasions a sensation as if the bowel were filled causing a frequent desire to evacuate; the effort is strained and painful.

The abdominal walls are usually retracted and the back arched. The pulse is not materially changed excepting in extreme cases. Great frequency of the pulse denotes gravity and danger, but the reverse does not hold good as sometimes in fatal cases it is but little quickened.

Fever is exceptional although at times it may run very high. The nose is hot and dry, the tongue often coated. Thirst is a prominent symptom. Vomiting may occur and a greenish matter be expelled. The loss of strength varies according to the intensity of the intestinal inflammation. In extreme cases running to a fatal termination, the discharges become putrid, the breath offensive, the respiration more rapid, the eyes sunken, the expression pinched and anxious. The pulse grows weak and feeble, evacuations occur more frequently, and at last are involuntary and beyond the control of the animal. Paralysis seems to invade the extremities, the skin becomes cold and clammy, the stench intolerable, finally death ensues from exhaustion.

Prognosis. —The disease intrinsically tends to recovery. It is a distressing affection, but properly treated need seldom result fatally, provided the system has not been weakened by some previous disease or some co-existing derangement. Exceptionally dysentery eventuates in the chronic form of the disorder.

Treatment. —It is desirable that as early as possible the contents of the intestines, and more especially the larger, be effectually removed. Nature evidently endeavors to relieve the bowels by a diarrhoea which precedes the dysenteric discharges. The treatment should be commenced with an effective purgative; castor oil is the more appropriate remedy and a tablespoonful should be given. Following the oil, opium must be relied upon. The following mixture is recommended. —

> R   Morph. Sulph.　　　gr.ij
> 　　Sodæ Sulphatis　　　℥iij
> 　　Aquæ Cinnamon.　　℥iij

Ft. Mist. Sig. Dose one teaspoonful every five or six hours.

It will be necessary in very many cases to repeat the castor oil especially if fever manifests itself, or the discharges assume an offensive odor. The writer has had too little success in the use of the so-called astringents to advise their administration in the early stage at least. Where the discharges assume a character near the normal but watery, then the following mixture may be given. —

> R   Misturæ Cretæ　　℥iijss
> 　　Tinct. Catechu　　℥ss

Ft. Mist. Sig. Dose a dessertspoonful every three or four hours.

Vomiting sometimes occurs as a prominent symptom at times early in the disease, preventing the administration of medicine by the mouth. In such a case morphine can be given subcutaneously; or thirty drops of laudanum in starch-water be injected into the rectum. These methods need only be employed while the stomach is too irritable to bear what enters it.

The food should be bland and easily digestible; rather withhold it entirely than unwisely select that which is liable to be vomited, or add to the irritation.

Milk and lime-water in small quantities given frequently, will be quite sufficient at first. This can be varied by giving alternately broths, into which a raw egg has been broken. When any diet but milk is allowed some form of pepsine should be administered with the food; of the French preparation, five grains is the dose, of the American or saccharated, from five to ten grains.

The elixir bark pepsine and bismuth is an admirable combination and can be given in teaspoonful doses three or four times a day. If the symptoms indicate failure of the vital powers, concentrated nourishment and possibly stimulants will need to be given freely.

When the disease is under control and convalescence commenced, tonics should be employed.

The beef wine and iron at first in dessertspoonful doses four times daily, then the elixir calisaya bark iron and strychnine, one teaspoonful twice a day. The pepsine should be persisted in until recovery is complete.

CONSTIPATION.

The term constipation and costiveness have the same significance, and denote insufficient evacuation of the bowels. Constipation exists as a functional disorder, and occurs as a symptom in various diseases. It is extremely frequent among dogs, and while alone it might not be considered as a serious affection. It demands consideration from its importance as giving rise to other ailments, and the difficulty experienced in attempting a cure unless its causes are correctly interpreted, and treatment is judiciously administered.

The affection may be occasional and of short duration, or it may be habitual.

Causation.—The causes of this disorder and the circumstances which may contribute to it are very many. Mechanically the abdominal muscles render important aid in the operation of unloading the bowels. These muscles become weakened by distension as in pregnant bitches, and in animals over-fed and excessively fat.

Lazy, sleepy, indoor pets are supposed to be especially prone to constipation, but it is probable other causes are more active in its production; among them excessive neatness and restraint, the caprices of their owner's, rather than their own wants, directing when they be allowed out of doors. If nature's calls or promptings are continually disregarded or resisted, the sensibility of the lower bowel, or temporary depot for the excretion becomes lessened, the accumulation goes on and paralysis of the parts follow repeated and continued distension of the intestine.

Food, highly nutritious and digestible, leaving but little residue to be thrown out, contributes to constipation. Active exercise may induce the affection by consumption of the fluids of the body. If deprived of drinking water, on the same principle the discharges are dry and hard.

A uniform, unvaried diet taken day after day, tends to obstruct the bowels. Again the habitual feeding of very coarse food induces the disorder, the intestinal activity becoming lessened, fatigued as it were, by the continued strain or demand made to throw off so large a residue.

In the constipation occurring in the course of chronic diseases, certain factors show in its production; among the many may be mentioned defective nutrition, degenerative changes in the muscular coat of the intestines, chronic catarrh, and deficient secretion of digestive fluids.

Symptoms. —The discharges are dry, hard, and lumpy, passed with difficulty, and often severe pain. The operations in many instances are incomplete and frequent attempts are required to effect relief. The straining efforts of the animal at times, causes an eversion of the bowel and its protrusion through the anus. Associated with these symptoms are often observed vomiting, offensive breath, loss of appetite, a congested appearance of the eyes, coated tongue, and a dull, heavy, listless manner.

Treatment. —Occasional and but slight constipation is readily relieved. Dietetic means should be first employed, and if insufficient, medicinal remedies can be resorted to.

Dogs returning from bench shows almost always suffer from this derangement, and every trainer and care-taker have their own peculiar nostrums, with which the animals are dosed after exhibitions, whether the need of medicine exists or not.

The writer has liver in quantity boiled the day previous to the return of his dogs from exhibitions. This is mixed quite freely with their first meal, the effect watched, and the feeding of it continued until the need is no longer evident. This method will prove quite sufficient in nearly all cases.

If the constipation be more than slight, and yet no marked discomfort is apparent, the syrup of buckthorn or calcined magnesia may be added to the food twice daily until the bowels move freely.

The treatment of habitual constipation requires judgement and perseverance. The means which may be employed are various. In commencing treatment the object is to completely unload the bowel. If the animal shows by his distress and constant efforts a hardened mass lodged in the rectum, local treatment must be employed, and mechanical means resorted to. Warm soap-suds should be freely injected, an interval allowed for its return, and then another injection be given. If after repeated efforts this is not effectual, several syringes full of sweet or linseed oil should be thrown up and another interval allowed. If that does not suffice, then the finger must be well oiled and introduced into the rectum and the mass be broken down, thereby insuring its passage.

A radical change should then be made in the diet. Meat should be well cooked with vegetables, the latter when well done to be crushed finely and returned to the kettle; then with corn meal, wheat middlings, or oatmeal be mixed and cooked thoroughly. These articles with their indigestible constituents will slightly irritate the entire canal through which the food passes, thereby stimulating it to increased activity. This means must not be persisted in indefinitely, lest the digestive organs become fatigued from the continued strain forced upon them.

If the animal is poor in flesh, the addition of cod-liver oil to the food will act beneficially.

If medicinal remedies are demanded, this general rule is a wise one: better give laxatives twice or three times daily in small doses, rather than give the same in one large dose. Probably no better laxative can be selected for repeated administration than the syrup of buckthorn, which for a short time might be added to each feeding in one or two teaspoonful doses as demanded.

Laxative remedies, when the need of their persistent use is indicated, should have some tonic added to counteract their depressing effect. When it is remembered that habitual constipation in a measure results from a paralysis of the lower bowel, the use of nux vomica and belladonna can be appreciated. Both agents with colocynth are ingredients of the following pill.—

> ℞   Ext. Colocynth. Comp.   ℥ss
>   Ext. Nucis Vomicæ   gr.iv
>   Ext. Belladonnae   gr.ij
> Ft. Pil. No. xx. Sig. Dose one pill at night.

Sufficient exercise must be allowed, and this additional fact be remembered that the mind has a powerful influence over intestinal activity. Chain up for a short time each day a perfectly healthy dog and derangement will follow; not from the want of exercise, but from the mental influence of the restraint, as he frets, feels neglected and grows surly constipation will be observed.

ENTERITIS.

Enteritis is designated by nearly all writers as inflammation of the bowels. The name properly signifies an inflammation of the intestine, restricted in its application to the small intestine. The disease attacks the mucous membrane, involves the walls of the intestines, and sometimes extends beyond to the tissues in the part where the inflammatory process is going on.

Causation.—The cause is not always appreciable. The intestinal mucous membrane is especially liable to inflammations, and even slight irritations may suffice to excite the same.

Improper food; traumatic injuries; exhaustion from over-work; exposure to cold in damp draughty kennels; abuse of cathartics; irritant poisons; these are the more common causes. The presence of worms, obstruction, liver derangements, prolonged suffering from colic are influences which may induce the affection.

In pulmonary and certain other chronic diseases, enteritis is frequently associated. An intestinal complication is not uncommon in distemper. Certain climatic changes are well known to influence the frequency of the disease.

Symptoms.—The symptoms of enteritis are not very characteristic, at least they vary considerably in their degree of development in different cases. Constipation usually exists at first; then follows the most frequent symptom of intestinal inflammation, diarrhoea, the discharges being thin and watery mixed with mucus. Blood is a rare admixture, excepting in cases where the rectum becomes inflamed either by the extension of the disease from the small intestine, or by the passage out of acrid and irritating discharges.

In many attacks the disease is ushered in by a chill as shown by shiverings, then the usual constitutional signs of inflammation follow. The nose becomes hot and dry, the tongue parched, the eyes reddened, the urine scanty, and the pulse rapid and bounding.

Abdominal pain is an almost constant symptom. Its character varies considerably; sometimes it is colicky, having intervals of increased severity; at others it is continuous, dull, aching, and aggravated by pressure.

The expression of the animal denotes anxiety, his moanings and restlessness, his discomfort. When standing his body is arched and his tail is pressed tightly between his legs. The abdominal walls are firm, hard, and tense. Vomiting is quite a common symptom, sympathetic in character generally, rather than due to any gastric irritation. The disease may extend and involve the stomach.

The breathing in severe cases is hurried and painful; the animal seems to avoid a deep inspiration. By an examination of the anus it will in many cases be found that the parts are reddened, hot, and dry.

In cases running a fatal course the strength suffers severely. The breathing grows more hurried, the pulse more feeble; the body emits a sickening odor; the signs of complete exhaustion follow rapidly; the skin becomes cold and clammy; convulsions occur and the animal dies.

Diagnosis. — This should not be difficult. The disease is to be discriminated from inflammation of the stomach, colic, and dysentery. Vomiting may suggest gastritis, but the symptoms of that disease are in the earlier stages of a much graver character than those of enteritis, and point clearly to the stomach. The two diseases may be combined, as is the case of irritant poisoning.

Colic is a functional trouble in which the pain is intense, without fever or other inflammatory signs, and with it constipation is more often associated than diarrhoea. The absence of the characteristic dysenteric discharges and the accompanying straining, should readily exclude dysentery. The only room for doubt is when an inflammation of the large intestine is superadded to enteritis.

Prognosis. — In inflammation of the intestines the prognosis varies much, according to the severity of the disease, the cause which produces or maintains it, the general strength of the animal, etc. If not associated with any other affection, recovery often takes place. The disease is however a grave one and danger invariably attends.

Treatment. — Like the prognosis, the treatment also is determined by the exciting cause of the attack. Naturally the first step if the animal suffers great pain, is to control it with opiates. One of the most important measures next to be employed is the removal of the intestinal contents. This precaution should never be neglected, even when diarrhoea is present. The writer is well aware that authors consider the employment of purgatives in this disease contra-indicated, and one strenuously protests against their use "as almost sinful." Why such a radical theory cannot be accepted may be better understood after considering certain facts.

The important end in the treatment of all inflammations is quietude of the inflamed part; this cannot be secured in inflammations of the intestines until the contents are effectually removed, thereby preventing their continued passage over the inflamed surface. Again when the contents of the intestines are indigestible and irritating, nature's own prompting and remedy is a diarrhoea. Can we do better than imitate nature? When the beginning of the large intestine is the seat of the disease, the use of cathartics is attended with some risk, but that portion is rarely seriously involved and little or no danger exists from them at the onset of the inflammation. After an enteritis has been present for a time and the tissues in the neighborhood have taken on the inflammatory process then a cathartic would be ill-advised. After a consideration of these facts it seems fair and rational to draw this conclusion, namely: the evacuating treatment is judicious and appropriate in the early stage of enteritis and should never be neglected.

If the attack has been preceded by constipation, or there is reason to suppose the lower bowel is loaded, an injection of oil or warm soap-suds may be given; in fact that intestine should be thoroughly evacuated by this means. Then instead of giving a large purgative dose of castor-oil, it were better to administer it in divided doses. Two tablespoonfuls of the oil should be beaten up with two raw eggs and of the mixture, a tablespoonful be given every two hours until the bowels move freely; at the same time employing opiates to control the pain; also adding to the relief of the animal by making hot applications to the abdomen.

A large piece of old blanket wrung out in hot rum and water, equal parts, applied to the bowels and changed frequently, will be found to aid greatly in relieving the distress.

The selection of a suitable diet is of the greatest importance in treatment. Only those articles easily digested should be given. Milk, raw eggs, scraped beef, broths, gelatine, teas of flaxseed and slippery elm are the more appropriate.

Of all medicinal remedies some form of opium is the most important, securing thereby rest to the inflamed parts. Of laudanum the dose should be fifteen drops; of paregoric one teaspoonful; of morphine one twelfth of a grain. Either of these preparations can be given singly in the early stage of the disease, and be repeated from two to four hours, as is necessary to subdue the pain. In the later stage it may be well to use the opiate in combination with an astringent as follows.—

$$\text{R} \quad \text{Morphiæ Acet.} \qquad \text{gr.ij}$$
$$\text{Acidi Tannici} \qquad \text{₃ss}$$
$$\text{Aquæ Camphoræ} \qquad \text{₃iij}$$

Ft. Mist. Sig. Dose one teaspoonful from two to four hours.

During convalescence the diet should be restricted for a long time, and when allowed to be more generous, the elixir bark pepsine and bismuth in teaspoonful doses may be given three times a day with the food.

INTESTINAL OBSTRUCTION.

Obstruction is possible in any part of the intestinal canal from the mouth to the rectum. The condition is an uncommon one, and when existing it is rarely recognized in time but is mistaken for some other disease of the the bowels.

Causation. — It may arise from many causes; more prominent among them are foreign bodies such as bones, wood, grass and other indigestible materials taken into the stomach; obstinate constipation, fecal masses becoming lodged in the bowels; pressure upon the intestines from without, as in cases of tumors, or the abdominal organs greatly enlarged by disease; solid growths within the intestine; the imprisonment of the bowel in holes, fissures and hernial rings; strangulation by false ligaments, or bands of lymph; twisting, and the formation of knots in the intestine; obstruction by the bowel itself as in intussusception, a condition where one portion of the intestine falls into another and becomes strangulated.

Symptoms. — The selection of a typical case of acute internal strangulation for an illustration can alone convey an adequate idea of the symptoms.

Pain is usually the first manifestation sufficient to attract attention, although in some instances an insignificant diarrhoea or constipation may have previously existed. The pain is severe, colicky in character and recurs at intervals; the symptoms at this stage are identically those described in colic, for which obstruction is commonly mistaken and treated.

For convenience it is presumed the usual domestic remedies have been employed, and under the impression a free passage of the bowels would bring relief, not only a cathartic but an injection has been administered. But success does not attend the use of these remedies; the bowels do not move, or at least only slightly, the pain persists in all its intensity, and vomiting becomes frequent. The abdomen becomes distended, the expression piteous and anxious, the eyes congested and sunken. The respiration is superficial and frequent, the pulse small and rapid. The animal makes frequent attempts to empty the bowels; these efforts are painful and increased by the failure to purge. The vomited matter is composed, first, of the contents of the stomach and then of greenish matter, later a dirty green, then brownish having something the appearance of diarrhoeal discharges; finally if life is prolonged, matter which should naturally have been thrown off by the bowels appears mingled with that vomited, accompanied with the characteristic fecal odor At this time the animal is in a condition of collapse, the skin cold and clammy, vomiting frequent, breathing rapid, thirst great, pain exhausting, eyes leaden, tongue dry and covered with a dirty brown coat, pulse thin and thready soon is no longer felt, and the animal dies.

This is a picture of a typical case of intestinal obstruction running an acute course. It will be appreciated that few cases have symptomatic phenomena in common. In many instances they run a course which might not improperly be termed chronic; the symptoms then are vague and ill defined. This is commonly the case when the obstruction is due to the impaction of foreign bodies such as woody fibres, bones, etc.

Some cases of the so called chronic affection have been reported throughout which aside from discomfort, loss of appetite, obstinate constipation, and some fever late in the attack, no positive symptoms were present absolutely indicative of the trouble.

Diagnosis.—The diagnosis is usually made at the autopsy, not before. Where the attack takes on an acute course, it might be mistaken for a case of poisoning, more especially with arsenic; but a careful consideration of the symptoms should obviate the danger of confusion. Generally the difficulty in diagnosis will be to discriminate obstruction from colic and acute peritonitis. The persistency of pain, the appearance of tenderness and vomiting, with such marked constitutional symptoms will soon point to a graver affection than colic.

In acute peritonitis the symptoms develop more gradually; the tenderness is more diffused over the whole abdomen, the muscles of which are more rigid, and the matter vomited is different in character. It should also be remembered that peritonitis more commonly follows traumatic injuries and an absence of such a history would weigh considerable.

In obstruction often a lump or tumor appears in the abdomen, showing the locality of the stoppage. At best a diagnosis will be difficult; the considerations noted will aid some in differention.

Prognosis.—If the obstruction is caused by obstinate constipation and impaction of foreign bodies, the chances of recovery are greater. When the intestine is closed by a twist it might straighten itself, a knot might loosen, a strangulated loop might become free, and an intussusception become disengaged; again that portion of the intestine which may have invaginated or entered into another, might slough off and pass out by the bowels. Such happy results are however very rare, and a fatal termination in the majority of cases may be anticipated.

Treatment.—Theory suggests many methods of treatment; experience tells how futile they have proved. The indications are to secure as near perfect rest of the intestinal canal as possible, to relieve pain and support the powers of life. Cathartics must never be given, as they interfere with the objects just mentioned and are deadly in their effect. Opium should be given in sufficient quantity to control the pain, and the strength should be maintained by concentrated and nutritious food with stimulants if needed.

PERITONITIS.

The peritoneum is a serous membrane partially investing all the organs in the abdominal cavity. An inflammation of this membrane or peritonitis may be either acute, circumscribed, or chronic as regards the degree and duration of the inflammation.

The disease is an exceedingly grave one and fortunately but rarely met with among dogs.

Causation.— Traumatic injuries such as kicks and blows, exposure to cold, and whelping. In the great majority of cases the disease is incidental to other affections of the abdominal organs. In intestinal diseases the inflammatory process may extend and involve the peritoneum, or ulceration may induce it through perforation.

Symptoms.—In acute inflammation of the peritoneum the symptoms may be readily recognized. The pain is agonizing; to this the sharp shrill cries of the animal fully testify. While he has strength to remain on the feet his restlessness is ceaseless, and in no position can he find relief. He breathes solely with the chest, the abdominal muscles being rigid and fixed. The expression is anxious, the eyes reddened and sunken; the pulse small, wiry, and resisting; the tongue dry, vomiting constant; constipation is obstinate. The abdomen is distended, tense, and great tenderness is more or less widely diffused. If the case proceeds to a fatal termination the abdomen swells with an effusion, the pulse becomes quicker and weaker; paralysis follows; then exhaustion and death.

Diagnosis. — In severe forms of the disease under consideration, this should not be difficult. From enteritis it is distinguished by greater pain, greater abdominal tenderness and distension, absence of diarrhoea, and the presence of evidences of a much graver disease. Colic may be eliminated on the same points of difference between that disease and enteritis.

Prognosis.—General peritonitis is an exceedingly grave disease from which there is little hope of recovery. The danger is greatly increased by perforation and co-existing diseases.

The affection may run rapidly to a fatal termination, destroying life within a few hours, or the issue may be prolonged several days. Recovery might take place from the acute form of the disease, but such a result must be painfully rare.

Treatment. — Opium is the sovereign remedy and on that reliance must be placed. How large a dose and how often, it is difficult to direct. The natures of animals vary; some are susceptible to a small quantity, others bear large doses of opiates with but little effect. Again the pain may be agonizing in some cases and less severe in others. Still in other attacks vomiting will be so persistent that medicine will not be retained, in which event morphine must be administered subcutaneously in one eighth grain doses. Where the medicine is not rejected laudanum is a convenient preparation of opium, and should be given in twenty drop doses. Should the animal show no signs of sleeping and the pain persists, doubtless the dose could be safely repeated every hour and a half or two hours until some relief is secured. Although obstinate constipation exists, purgatives must not under any consideration be given during the acute stage. Remember the costiveness arises from an inflammation of the bowels, and they being unable to expel their contents.

Hot applications to the abdomen are indicated, and it matters little their character as long as they are hot. Nourishment should be given ice cold, the most concentrated being demanded.

Partial or circumscribed peritonitis is as the name implies an inflammation limited to a portion of the peritoneum. It is a complication of a previous affection of the parts covered by the inflamed membrane. Where this inflammation exists, a disease of the tissues beneath may be inferred.

Chronic peritonitis is less common than the acute. Occasionally the latter eventuates in the chronic form, but rarely so; it is more often associated with some diseases of the abdominal organs, previously existing or present. As chronic peritonitis must sooner or later eventuate in death, as it is insidious and its existence is rarely detected except at the autopsy, an extended consideration is an absurdity.

$\mathcal{P}ILES.$

Hæmorrhoids or piles are small tumors situated near the anus. They consist of folds of the mucous and sub-mucous tissues, and usually contain large veins. These tumors may be in a state of congestion, swollen from inflammation and very sensitive, or they may exist as a simple thickening of the parts and comparatively free from pain. Sometimes the veins supplying the anus become morbidly dilated forming knots, and at times the blood in these distended vessels coagulates or clots, and a solid tumor is formed. Again the whole mucous membrane of the lower part of the rectum may become swollen, sensitive, and protrude at every evacuation of the bowels, causing great pain, and at times bleeding.

The tumors may be pendulous, varying in size, and hanging down from the rectum productive of great tenderness and discomfort. External piles may be met with in round bunches at the anus, in part covered by the mucous membrane and partly by the skin.

Causation. —The predisposing causes are any influences which produce a fulness of the abdominal blood vessels, or obstruct the return of blood from the rectum to general circulation. Among them may be mentioned diseases of the liver, insufficient exercise, pregnancy, and habitual constipation. When the latter complaint induces piles, they may be as much attributed to the irritation and congestion which it excites as to the impediment to the course of blood produced, for otherwise the tumors would disappear when the bowels were freely moved, which is not the case.

The exciting causes may be straining to discharge the bowels as in dysentery, violent purgative medicines, in fact anything that irritates the lower bowel.

Symptoms. —The dog will generally by his manner draw attention to the affection. Pain will lead him to lick the parts, and to relieve the itching which is commonly associated he will drag the anus along the floor, or will rub himself astride a bar or against a post. Swelling of the parts indicate the affection. Some blood may be found in the discharges.

Treatment. —The primary object is to remove the predisposing and exciting causes. If the animal is over-fed and has insufficient exercise, his diet should be restricted and work be insisted upon.

The bowels should be regulated and the discharges kept soft and copious. The domestic remedy, cream of tartar, and sulphur will probably be sufficient, and may be given each morning with the food, two teaspoonfuls of the former to one half a teaspoonful of the latter. If the piles are inflamed bathing them with cold water frequently will afford relief, and insure perfect cleanliness, one of the essentials. The following ointment should be freely applied to the swollen parts several times daily.—

$$\begin{array}{lll}
\text{R} & \text{Acidi Tannici} & \text{Ʒi} \\
& \text{Morph. Sulphatis} & \text{gr.iv} \\
& \text{Pulv. Camphoræ} & \text{Ʒss} \\
& \text{Ungt. Stramonii} & \text{Ʒi}
\end{array}$$

Ft. Ungt. Sig. External.

After the bowels have been regulated and the sulphur and cream of tartar are discontinued, it is advisable to give two teaspoonfuls of pure glycerin in each feeding for several weeks or until a cure is complete.

Tumors of long standing and not yielding readily to treatment should be removed by the actual cautery.

Hemorrhage from the rectum while it is frequently associated with piles, is rarely important enough to need interference; unless the loss of blood be sufficient to weaken, it will be a relief to the animal and no effort should be made to stop it.

If the bleeding is considerable and debility evidently results from it, a tonic of iron is demanded and may be given as follows.—

$$\begin{array}{lll}
\text{R} & \text{Ferri Sulph.} & \text{Ʒss} \\
& \text{Acid. Sulph. dil.} & \text{Ʒi} \\
& \text{Aquæ} \qquad \text{ad.} & \text{Ʒiv}
\end{array}$$

Ft. Mist. Sig. Dose one teaspoonful in a little water three times a day.

FISTULA IN ANO.

Fistula in Ano signifies an ulcer through the rectum, and a passage by the side of it down through the fibres of the sphincter ani, the muscle which surrounds the anus closing it by contraction.

Fistula of the anus may be met with in dogs, more commonly among the house pet overfed and deprived of needed exercise. This affection is a common result of abscess near the rectum. One eminent surgeon maintains "that it always commences with an ulceration of the mucous membranes of the rectum, and an escape of the fecal matter into the cellular tissue, which gives rise to abscess and fistula."

This affection may exist as follows — complete, having an external opening near the anus, and another into the bowel above the sphincter muscle — as a blind external fistula which has no opening into the bowel — as the blind intestinal fistula which opens into the bowel but not externally.

Symptoms. — Irritation and pain in the affected part causes the animal to act much the same as when suffering from piles, licking the anus and dragging himself along the floor, etc. If the fistula opens externally, less difficulty will be experienced in making a diagnosis; still in but few cases will detection be easy, but rather the reverse, as in many instances the opening will be minute and need close scrutiny to discover it. The course of an external fistula is that of a frequent recurring abscess; the cavity fills up and discharges, the opening then closes and again the cavity fills. In searching for a fistulous opening, a minute drop of matter in the centre of a slight swelling will often mark its location. To explore its track a small steel knitting needle will be sufficient.

When a blind internal fistula exists its presence may be suspected by a discharge, bloody, watery in character, and at times offensive.

Treatment. — The constitutional treatment may be left to owners and care-takers but the local surgical treatment never. The constitutional treatment is symptomatic; correct abuses, regulate the diet, and if weak, strengthen. Constipation should be overcome by judicious dieting; among other articles of food no better laxative is known than liver, which can be fed in sufficient quantity to keep the bowels regular, or if preferred the cream of tartar and sulphur as advised in piles may be given instead.

Exercise should be enforced. If tonic remedies are demanded, cod-liver oil, iron, or quinine may be administered.

A radical cure of fistula demands a surgical operation. Of several popular methods, the elastic ligature is advised. When an operation is imperative employ a competent surgeon and leave the method to his selection.

PROLAPSUS ANI.

The name denotes an eversion of the lower portion of the rectum, and its protrusion through the anus. The affection is more common in old dogs, but may appear at any age, caused by a natural laxity of the parts or by straining from constipation.

Treatment. — Whenever the protrusion occurs, the parts should be replaced as soon as possible after being carefully washed. If any difficulty is experienced, the fore finger should be well oiled and pushed up into the anus carrying the protruded part with it.

To radically cure the affection the bowels should be regulated as in piles, and tonics be given. The tincture of chloride of iron is preferred in ten drop doses well diluted and administered three times a day. In the severest cases use the following.—

$$\text{R} \quad \begin{array}{ll} \text{Ferri Persulph.} & \text{℥ ss} \\ \text{Pulv. Opii} & \text{gr.xv} \\ \text{Camphoræ} & \text{℥ ss} \\ \text{Ol. Theobromæ} & \text{q. s.} \end{array}$$

Ft. Suppositories No. xij. Sig. Introduce one into the rectum twice daily.

CHAPTER VII.

DISEASES AFFECTING

THE

SOLID ORGANS OF THE ABDOMEN.

ACUTE HEPATITIS.

Of the disease of the solid abdominal organs, the greater number and the more important are situated in the liver. Inflammation of this structure is termed hepatitis and may be either acute or chronic. These two forms claim separate consideration.

Acute hepatitis is an affection rarely seen among dogs in the colder climates. It is reported that the disease is not infrequent in the tropics; the course it runs there is however somewhat different, it being circumscribed and confined to a part of the organs, while here the inflammation generally acts upon the whole or greater part of the structure.

Causation. — Acute hepatitis in the cold climate is probably more often induced traumatically, by such influences as kicks and blows.

Various causes have been assigned by different writers, such as excessive use of purgatives, emetics, and exposure to cold and wet. The disease has been often known to follow severe cases of dysentery. Gastric inflammation may be the exciting cause, by interfering with the circulation in the liver. Acute hepatitis may be associated with other diseases, such as pneumonia, pleurisy and distemper.

The existing causes in many instances must be problematical; the subject requires much light before the direct influences can be determined.

Symptoms. — At times the symptoms are vague, and point to other organs rather than to the liver as the seat of the disease. Persistent vomiting may mislead and acute gastritis be suspected.

The attack is very often ushered in by a chill as shown by shivering, but this symptom is liable to be overlooked. The manner of the dog changes materially soon after the invasion of the disease, becoming dull and listless. His appetite usually at once disappears, and he manifests an appearance of discomfort, very marked in a proportion of cases, from which it might be inferred he was suffering from pain. His position assumed on lying down is somewhat characteristic, it being on his chest and belly. He shows a disinclination to exertion and on getting up his movements appear stiff and painful. Thirst is usually excessive; vomiting attends a certain proportion of cases and the odor of the breath becomes markedly offensive. The condition of the bowels is not indicative; looseness occurs in some, and in other cases the bowels are constipated; the two may alternate.

Unless the attack be associated with pulmonary disease, the respiration is at first unaffected. Fever is a constant symptom; the pulse becomes full, bounding, and rapid. Tenderness on pressure over the liver is characteristic.

Jaundice appears quite early in the disease, generally within four days after the first symptoms are manifested. The skin becomes yellow, the eyes of the same tint, or dull and lustreless; the mucous membrane of the mouth paler, the urine darker in color leaving an indelible stain, and the discharges are "clay colored." The pulse previously rapid now falls, sometimes below the normal. The enlargement of the liver or a swelling in some part of it now very likely appears, but its increase in size may not in all cases be sufficient to be appreciated.

The course of the disease is usually rapid, and may terminate in abscess, recovery without abscess, or resolve into the chronic form. When death occurs from abscess, a swelling over the region of the liver can usually be determined. Emaciation is rapid and associated with the usual signs of failure and exhaustion. The breathing changes, becoming quick and hurried; the pulse weak and feeble. The bowels assume the appearance of pregnancy.

Diagnosis. — Early in the disease the diagnosis is exceedingly difficult in many cases; in none will it be easy. The manner of the attack, the loss of appetite, the high fever, the local tenderness and evident pain, and the occurrence of jaundice, render the existence of acute hepatitis highly probable. If in addition to these symptoms the liver is found to be enlarged, the diagnosis may be made with reasonable certainty.

Prognosis. — Acute hepatitis is a disease attended with great danger, and will in nearly all cases prove fatal if an abscess forms. When associated with other diseases the danger is intensified. Even if recovery takes place convalescence is apt to be slow and tedious.

Treatment. — The purposes of treatment are to arrest inflammation, thereby preventing the formation of abscess.

The amount of pain and constitutional disturbance will indicate the use of opium, which should be given in the form of Dover's powder; dose five grains every six or eight hours. If a correct diagnosis of the disease could be made within a few hours of its invasion, the better treatment would be to give the tartrate of antimony and potassa, in one sixteenth grain doses every two hours for two or three days, when there is much excitement of circulation, a full, bounding pulse, and much fever. Unfortunately however, in many cases several days will elapse before a diagnosis is verified, then it is too late to use the drug in question.

Mercury is supposed to increase the secretion of bile, and did we believe this to be so, its use in this disease would be of doubtful propriety. The doctrine that this agent acts in the manner supposed is open to distrust. Of the modus operandi of mercury we know nothing, except that it probably acts through the medium of circulation, and it exerts a peculiar power which enables it to subvert diseased actions.

Where blood-letting is generally inadmissible, one of the measures most to be relied upon, is the very cautious employment of mercury. It is therefore advised that the following be given.—

R Pil. Hydrarg.　gr.v
Pulv. Ipecac.　gr.ij
Ft. Pil. No. xij.　Dose one four times daily.

These pills should be persisted in unless some indication presents to contra-indicate their use, or an abscess begins to form. If the bowels remain constipated, one or two grains of the extract of colocynth can be added to each pill.

Counter-irritation should be applied over the region of the liver, a strong mustard paste used first, and after the skin has become well reddened, hot poultices of flaxseed meal should be substituted.

After the inflammation has subsided two grains of the extract of taraxicum can be added to each pill. The system must be sustained from the first by a nutritious and concentrated diet, and a careful avoidance of all fatty food.

CHRONIC HEPATITIS.

It is stated by some that chronic hepatitis may be a sequel of acute inflammation of the liver. This seems probable, and yet from the peculiar nature of the disorder, in but few cases can the point of departure from the normal to the morbid changes be traced with sufficient accuracy to assign a cause beyond a reasonable doubt.

Symptoms. — The disease as a rule gives rise to few or no symptoms which point to the liver as the seat of disease, prior to the occurrence of abdominal enlargement. Preceding this manifestation the appetite is usually lessened, the manner of the animal often dull and heavy, and emaciation is progressive. As loss of weight persists the abdomen becomes distended, rendering the general appearance highly characteristic of the disease.

In the majority of cases the tongue becomes white, and the mucous membrane of the mouth and lips loses its healthy color, becoming pale and yellowish. Jaundice in a certain degree occurs in some cases; it is however rarely very marked.

The breath is usually offensive; the eyes are dull and heavy. Vomiting is quite frequent, the matter ejected often greenish. The urine is scanty and high colored, and the bowels constipated. As emaciation progresses the skin becomes thickened, rough, and scaly; the hair dry and brashy. Until the abdominal distension is sufficient to mechanically interfere with respiration, the breathing is unchanged. The pulse varies but little until late in the disease.

In certain cases instead of enlargement of the liver it may be reduced in size; the distension which then occurs is due to abdominal dropsy.

Diagnosis. — It ought not be difficult to make a correct diagnosis after important changes commence in the liver. If that organ is enlarged its outlines may be traced by firm pressure with the hand, and the shape of the abdomen is characteristic. When the animal is on his feet the enlargement is carried higher, and there is not that pendulous or hanging down appearance noted in dropsy, the disease with which chronic hepatitis may be confounded. In the former affection the shape of the abdomen would not be retained when the animal lies down, while in the latter it would remain much the same.

In dropsy occurring independent of liver affections, the abdomen is rarely alone involved; other parts are soon affected by the disease. When jaundice appears, other symptoms being present, the diagnosis if previously doubtful ought to be made with reasonable certainty.

Prognosis. — The disease if extended is incurable. Something may be done in some cases to improve the animal and arrest the progress of the affection; but the result of efforts must generally prove unsatisfactory.

Treatment. —No specific treatment can be suggested. Remedies should be addressed to the general system rather than the local affection. Tonics and nutritious food are indicated to combat depressing influences, and symptoms of prominence demand individual treatment.

If the animal is debilitated, ferruginous tonics promise the better results; the tincture of the chloride, or the syrup of the iodide of iron may be given three times a day in fifteen drop doses.

Mercury is an agent which theory would suggest in chronic disorders of the liver. In this disease it should be given a fair trial, and the effects carefully watched. To obviate any depressing influence it may exert, iron may be combined and given as follows.—

$$\text{R} \quad \text{Mass. Hydrarg.} \qquad \text{gr.x}$$
$$\text{Ferri Sulph. Exsic. gr.xxx}$$
Ft. Pil. No. xxx. Sig. Dose one three times daily.

During the use of the pills all other medicine should be discontinued and the mercurial be given a fair trial. If the results are negative the treatment should not be prolonged.

When constipation exists a laxative rather than a cathartic should be given, but only when needed. Podophyllin is a remedy advised by some; it is inferred that their reasons of its use is some theory they entertain that it has a specific action on the liver. There do not appear sufficient grounds to affirm that the drug has any such special powers.

For the purpose of overcoming constipation the following pills are advised.—

$$\text{R} \quad \text{Extract. Aloes.} \qquad \text{Зss}$$
$$\text{Ext. Colchici Acet.} \qquad \text{gr.x}$$
$$\text{Pulv. Ipecac.} \qquad \text{gr.x}$$
$$\text{Ext. Taraxici} \qquad \text{Зi}$$
Ft. Pil. No. xx. Sig. Dose one at night.

FATTY LIVER.

The term is used to denote a morbid condition of the liver increased in size by the excessive accumulation of fat, sufficient to interfere with the healthy action of that organ. Observation has shown that the quantity of fat contained in the livers of dogs is influenced by the diet, and a large quantity may be accounted for by that which is taken in with the food. Fat is however found in the body independently of that which is introduced by the stomach as experimental observation has proved. Oily matters must essentially enter to some extent into the composition of the food, in order to maintain the animal in a good condition.

All flesh and vegetable substances contain more or less fats which are not transported into the body to be deposited there unchanged. On the contrary they are altered and used up in the processes of digestion and nutrition; while the fats which appear in the body are in great part of new formation, produced from materials derived from a variety of sources.

Fatty liver rarely if ever occurs as an independent affection and its importance as relates to disease has never been determined. The presence of fats in an abnormal quantity, in the secreting cells of the liver, might by their presence seriously interfere with circulation, and might in extreme cases be sufficient to obstruct the bile and occasion slight jaundice.

The cause of the disease under consideration is no doubt largely influenced by fatty food, conjoined with insufficient exercise; the method of the accumulation of oily matters in the liver being much the same as in the production of the "foie gras" of geese.

Insufficient material for the study of the disease when it occurs independently of other affections leaves but an imperfect knowledge of the symptoms.

Impaired digestion, vomiting, progressive emaciation, loss of strength, and slight jaundice later in the disease, are the common evidences of fatty liver. If the organ is much enlarged the fact can be determined, but the increase in some cases is insufficient to be detected, and it is to be remembered that a diminution in size may occur from the accumulation of fat in the secreting cells.

In the treatment abuses are to be corrected, over eating prevented, fatty food excluded from the diet, exercise insisted upon, and tonics should be given if the need is apparent.

That starchy food has certain fattening properties is well known; it is not certain however whether the constituents of the same are directly converted into fat, or are first taken up and distributed in the system, and afterwards supply the materials for its production.

JAUNDICE.

The presence of biliary coloring matter in sufficient quantity to give a yellow color to the skin, constitutes the diseased condition known as icterus or jaundice. Strictly it is never an individual affection, it being a symptom, and incidental to various disorders.

Jaundice often appears when its cause cannot be clearly determined; again, it is an important morbid condition, convenience in considering which demands special recognition and classification. In considering it as an individual disease, cases are to be excluded in which it occurs with an affection, the existence of which is clearly determined.

In acute and chronic hepatitis and certain constitutional diseases it is to be regarded merely as a symptom.

Causation. — In a great majority of cases jaundice depends on obstruction of the passage of the bile to the intestine.

From obvious facts it is reasoned that more commonly the bile is reabsorbed after the secretion of it has taken place. Defective secretion however, either with or without obstruction doubtless occurs, and the constituents of the bile accumulate in the blood if not thrown out by some unusual channel.

It is conjectured that more or less of those indefinite symptoms which are commonly included under the name "biliousness," are due to a deficient elimination of the bile or at least one of its elements.

The more apparent exciting causes are over fatigue, the indiscriminate use of powerful emetics and cathartics, indigestion, obstinate constipation, traumatic injuries, and sudden chills.

When the functions of the skin are arrested by cold there are two ways in which disease is supposed to be produced. One relates to the circulation; the blood being driven from the surface accumulates in the internal organs inducing congestion.

The other method of action relates to the increased activity demanded of certain organs, to compensate for the deficient elimination of the skin. It is to be remembered that the skin is an important medium, through which matters proper to be evacuated from the body are expelled. Again any material change in the condition of the skin affects the perspiration, the office of which is principally to regulate the temperature of the body.

Thus it will be seen that disease may not infrequently originate from the action of cold applied to the surface of the body, as in sudden immersions in water and too rapid cooling after a hard run; the radical and rapid changes of temperature being the most active.

Some authors include fighting and prolonged sport among the causes of jaundice. While the direct action of these influences is unexplained, it can however be appreciated, that in a variety of ways irritation and excitement can induce changes in the system, sufficient to give rise to the disease under consideration.

Symptoms. — The signs in the earlier stage are vague and confusing. The manner of the animal changes, he becoming dull, languid, and averse to exertion. The reabsorbed bile appears to exert a narcotic influence on the nervous system, producing dulness, a disposition to sleep, a fall in the pulse, and a torpid condition in the functions of the body generally. The nose, mouth, and breath are hot and dry; the abdomen hard, and back arched. The coat loses its glossy look and becomes dry and staring.

As the bile exerts its influence on circulation, the respirations become less frequent. In some cases a disposition to constantly scratch is noted. The appetite disappears, thirst becomes excessive, and vomiting frequent; the matter expelled being greenish and at times in extreme cases bloody. Pain is at times evident and colicky in character.

When tenderness and pain both exist in the region of the stomach, it is evidence of a low type of inflammation of the mucous membrane lining the stomach and upper part of the intestine.

Generally the bowels are constipated, but not always as diarrhœa sometimes occurs. The appearance of the discharges afford evidence as to whether the obstruction is complete or partial.

In the former the passages are clay-colored or ashy, while in the latter they are nearer the color in health. A peculiar fœtid odor is noted in discharges devoid of bile. The appearance of the yellow tint dispels all doubts as to diagnosis if the other symptoms have proved insufficient. The mucous membranes first show the coloring; the skin then is rapidly affected. The urine in jaundice is loaded with bile, and when voided leaves a decided stain.

Nutrition for a time may be but a little affected; then emaciation is rapid. When the disease progresses to a fatal termination the pulse falls, the extremities grow cold, the skin clammy, and death ensues.

Diagnosis. — The appearance of the yellowish tint renders the diagnosis positive. To determine the cause in individual cases is far from easy. The symptoms in the early stage of the attack, viz., loss of appetite, vomiting, and tenderness, indicate a possible inflammation of the mucous membrane of the stomach and upper part of the intestine. Sometimes gall-stones form in the gall-bladder, and in passing down into the intestine become lodged and obstruct the canal; pain of the severest character is the evidence of this accident.

Worms sometimes cause obstruction, but they with certain other causes can only be determined at the autopsy.

Prognosis. — The gravity as well as the symptoms depend on the morbid conditions which give rise to the affection. The discharging duct or canal leading from the gall-bladder enters into the intestine near the stomach; in sub-acute inflammation of the mucous membrane lining these parts, the irritation is liable to extend into the duct and cause its lining to become swollen, the passage closing and becoming obstructed, it being too small to admit of much swelling and remain open. The prognosis in such cases is favorable. If a gall stone becomes lodged, recovery will depend entirely upon its passage into the intestine or backward up into the gall-bladder.

When death occurs in jaundice from the retention of bile, the liver is the seat of serious disease. When permanent obstruction exists without serious structural changes elsewhere, life may be prolonged and nutrition sustained for a long time; sooner or later however, the vital powers fail, the body wastes, exhaustion follows, and death results. In severe cases terminating fatally the disease runs a rapid course, the duration generally being from three to five days.

Treatment. — It is to be remembered that in the majority of cases jaundice is due to a sub-acute inflammation of the stomach and adjoining portion of the intestine, and the natural tendency of the disease is to recovery. In diet and medication all influences calculated to add to, or prolong the irritation must be studiously withheld. Unless the urgency of symptoms demands a more vigorous treatment, a bland and digestible diet with gentle laxatives will suffice.

Calcined magnesia should be given in teaspoonful doses in milk two or three times daily to keep the bowels active. All fatty food should be withheld, and milk and gruels largely be depended upon. If pain exists opiates should be employed.

The attacks which occasionally occur resembling colic, are at times due to the formation of gas; by the absence of bile the contents of the intestines are no longer influenced by its antiseptic property, decompose and gas is generated.

In severe attacks of jaundice in which the constitutional symptoms indicate gravity, the need is urgent and the treatment must be energetic and immediately applied. The animal should be placed in a warm room, it being remembered the skin must be protected from cold, lest its eliminative functions be destroyed, and one avenue through which the bile can be thrown out of the system be closed. Calomel is the agent most to be depended upon, and should be given in doses of one half a grain each, four times a day. Rather than combine the calomel with laxatives it is wiser to give them separately and only as the need is apparent. Hot baths if needed will increase the activity of the skin, and sweet spirits of nitre in one half teaspoonful doses, given in water every two or three hours, stimulate the kidneys. Food should be milk, raw eggs, scraped beef, and broths.

TORPOR OF THE LIVER.

Functional disturbances of the liver are very common among dogs. The character of their food, irregularities in feeding, and insufficient exercise, promote a disorder which may be termed torpor of the liver, or as more commonly designated biliousness. Properly torpor in this instance means a deficient secretion of bile; still the term is often used in cases where the morbid condition is somewhat obscure. Over feeding, indigestible food, various irregularities, insufficient exercise, and over fatigue are among the more evident causes.

Symptoms. —Dulness of manner, offensive breath, capricious appetite, whitened tongue, and at times a hot nose are among the prominent symptoms. Constipation may exist for a time, to be followed by a diarrhœa. In the former instance the discharges are hard and sometimes clay-colored; in the latter loose, greenish, black or tarry in appearance.

The coat of the animal loses that bright characteristic gloss, becomes rougher and lustreless, and at times a skin eruption called eczema is asssociated.

An animal suffering from simple functional disorder of the liver, shows no very marked symptoms; still it is evident to an observer some ailment is depressing him. Under excitement his manner changes and nothing irregular is apparent, but when food is placed before him, or while in his kennel or about the house, he again becomes dull and listless.

Treatment. —Correcting all abuses are the first indications in treatment. Food should be allowed but twice a day and then the simplest diet be chosen. If the animal has been restrained, greater freedom should be allowed, and exercise if necessary be insisted upon.

Taraxicum or dandelion, it is generally accepted, has a specific action upon the liver, exciting it to secretion when torpid. One dose of this daily should be given, the freshly prepared extract preferred, and of it one half a teaspoonful is the proper quantity. After this has been administered, if the bowels are inclined to constipation, one half a teaspoonful of calcined magnesia may be given with the food, once or twice daily as needed.

PASSAGE OF GALL STONES.

The term calculus denotes a stone or gravel, or unorganized concretion found in the body, as in the bladder, gall-ducts, kidneys, etc. Biliary calculi are usually formed in the gall-bladder, and their passage down the duct or canal into the intestine is if they are large, attended with the severest pain. The reasons for their formation are not understood; it is presumed that certain constituents of the bile are deposited in consequence of the presence of mucus in the gall-bladder or ducts, as the result of inflammation. Other reasons are assigned, but sufficient evidence to substantiate these theories is yet to be secured.

Symptoms. — The presence of calculi in the gall-bladder cannot be determined until their passage out is commenced, and then only when they are of sufficient size to distend the walls of the canal. Pain is the prominent symptom and is of so violent a character in some cases the suffering is extreme. The occurrence of the pain is usually abrupt, no indications of ill health preceding it.

Vomiting sets in early and continues throughout the attack. The expression and manner of the animal is indicative of the great distress he is experiencing. The bowels are constipated, and in many cases if a movement occurs the discharges are clay-colored. Unless the attack is of long continuance the pulse shows but little change from the normal. The pain may be persistent without intermission, or it may take on a paroxysmal form, vary in intensity and finally end as abruptly as it first appeared.

Treatment. — Opium in full doses is the only agent to be relied upon. Astonishing quantities of this drug will be borne and little or no effect be observed. Of the solid opium the dose is one grain every two or three hours or until the pain is controlled. If vomiting contra-indicates its use by the mouth, one eighth of a grain of morphine should be given subcutaneously or double that dose by anal injection. If relief is not secured by this treatment the animal should be etherized and kept so for a time, then allowed to return to consciousness, and if the pain still persists the ether should be re-applied. It must be remembered that opiates need to be given with great care and not too often. While the pain is intense it antagonizes the narcotics and no harm results, but the gall-stone may suddenly pass from the duct into the intestine and then the pain will at once cease. If the system is too heavily loaded with opium, its antidote the pain being no longer present, the result is poisoning by the drug.

Various agents have been suggested to act as a solvent on the calculi and thereby prevent a recurrence of pain, which is almost certain to follow sooner or later, but it has not been proven that their use has resulted in marked success.

CANCER OF THE LIVER.

Cases are on record where the liver has been the seat of cancerous changes, proving the possibility of such an affection invading that organ. The disease in nearly all if not all cases, attacks other parts or organs of the body first, and the changes in the liver are of secondary occurrence.

It has been found circumscribed or confined to certain portions of the organ, and appeared in the form of nodules or tumors varying in size. Sometimes a tumor has been found to occur singly; again the whole liver has been studded with them.

Cancerous deposits take on two varieties which are termed the hard and the soft. With either the liver is usually enlarged, and when the former exists its presence can be more easily determined, the hard tumors being felt. The soft variety can rarely be detected during life.

Cancer of the liver rarely occurs except in very old dogs. If the bulging tumors can be felt, a correct diagnosis might be reached. The general manifestations of the disease are progressive emaciation, debility, abdominal enlargement, and possibly jaundice late in the affection.

Dropsy of the extremities is also liable to occur. Digestion is destroyed and the appetite lost. Diarrhœa usually occurs, and all the signs of exhaustion are present. Pain may be prominent as a symptom.

Treatment can have no other result than palliation, and possibly the prolongation of life.

AFFECTIONS OF THE SPLEEN.

The special function of the spleen has never been clearly determined; like other glands its action is to modify the constitution of the blood. The precise alteration which is effected by its passage through splenic tissue has not been discovered. The spleen may be removed from a dog without its loss producing any permanent injury. The experiment has been frequently performed, and among the most constant effects noted are increase of appetite and an unnatural ferocity of disposition.

The spleen cannot therefore be regarded as a single organ, but as associated with others which may completely or to a great extent perform its function after removal.

Acute inflammation of the spleen or splenitis is an exceedingly rare affection. The symptoms which have been observed are restlessness, tenderness over the organ, some fever, loss of appetite, vomiting, great thirst, and as Youatt observes, "shivering, the ears cold, the eyes unnaturally protuberant, the nostrils dilated, the flanks agitated, the respiration accelerated, and the mucous membrane pale." The same author mentions a discharge of a yellow, frothy mucus by vomiting. By this it might be implied that in those cases an abscess had formed in the spleen and the pus was expelled by vomiting.

The disease will rarely ever be diagnosed during life, and were it, the general principles of treatment would be much the same as in acute inflammations of other organs similar in structure.

In certain diseases of the general system, and in some affections of other organs the spleen becomes enlarged for a time and then returns to its normal size.

Chronic enlargement of this gland, unless it be greatly increased in size, is rarely attended with any symptoms to indicate the condition. Were it to be discovered and clearly diagnosed the treatment would be symptomatic, and the same as in chronic hepatitis.

The spleen takes on degenerative changes in common with the liver, but a consideration of them can be of no possible interest or profit to the general reader.

DISEASES OF THE PANCREAS.

While the fact is recognized that the pancreas may be the seat of inflammation as in the other glandular organs, the liver, kidneys, etc., it is also apparent a correct diagnosis of morbid changes must be so difficult it would be useless to discuss them at any length in this work. Of all organs similarly constructed probably the pancreas is the least liable to inflammations. Doubtless they take on much the same degenerative changes observed elsewhere, and are secondarily involved by the extension of diseases in adjacent parts.

A discharge of fat from the bowels is supposed to be a diagnostic symptom of pancreatic disorder, but there is no reliable evidence to sustain this opinion.

SUGAR IN THE LIVER.

· Besides the secretion of bile, the liver performs another important function, viz., the production of sugar. While a consideration of this subject is not essential to the doctrine of diseases, certain facts pertaining to it as discovered by M. Claude Bernard, may be of interest.

So far as is known the presence of sugar in the liver is common to all species of animals. The average percentage in the healthy liver of man and the dog is about the same.

. The system does not depend for its supply of sugar entirely upon external sources; saccharine matter is produced independently in the tissues of the liver, whatever may be the nature of the food subsisted upon. Bernard kept two dogs under his own observation, one for a period of three, the other of eight months, during which time they were confined strictly to a diet of animal food, boiled calves' head and tripe, and then killed. Upon an examination, the liver was found in each instance to contain a proportion of sugar fully equal to that present in the organ under ordinary circumstances.

In considering this sugar producing function, an important question naturally arises, bearing upon the proper feeding of dogs. It has been maintained by some that starchy and saccharine matters are utterly detestable to the canine race. A consideration of a few facts will render apparent that this extreme doctrine is untenable.

While it is certain starchy matters are not digested in the stomach, but pass unchanged into the small intestine, experiments have proved beyond all doubts that the intestinal fluids of a dog transform starch into sugar with the greatest promptitude, and it is then as rapidly absorbed. If a dog is fed on boiled starch and meat, while some of the latter remains in his stomach for eight, nine, or ten hours, the starch begins immediately to pass into the intestine, where it is at once converted into sugar and absorbed. The whole of the starch even may have completely disappeared in an hour's time. It is evident therefore that nature has made ample provision for the digestion of starchy food.

The writer maintains that while meat is the proper food for a dog, he as strongly insists that starchy matters in small quantities, far from being injurious, are really conducive to the health of the animal. Although the saccharine matter becomes changed after absorption, these same chemical changes themselves serve to maintain the integrity of the blood, and the healthy nutrition of the body.

CHAPTER VIII.

DISEASES

OF

THE URINARY ORGANS.

NEPHRITIS.

Nephritis or inflammation of the kidney is a disease rarely met with in dogs. Its infrequency is relatively the same as the occurrence of acute inflammations in other glandular organs.

Causation. — It is possible that nephritis may arise from the excessive dosing of certain drugs acting directly and powerfully on the kidneys, such as saltpetre, cantharides, and turpentine; the second used in blisters, the latter in the treatment of worms.

Injuries such as kicks or blows over the region of the kidneys, may induce the disease. Probably the more common cause is exposure to cold, or more especially the sudden cooling of an overheated body; shutting the dog into a cold kennel after a hard run, or allowing him to plunge into the water causes a violent contraction of the blood vessels of the surface, driving the blood to the internal organs, producing in them increased pressure and possibly congestion.

The use of certain ointments for mange, under some conditions may act upon the skin in a manner somewhat similar to cold, and thereby induce the affection.

The formation of stone in the kidney or renal calculi, might induce a disturbance and nephritis result.

Symptoms. — According to the severity of the disease the symptoms will vary greatly. Fever may be present in a noticeable degree in the commencement of the attack, and it may be absent.

When the skin is hot and the pulse runs high, vomiting, complete loss of appetite, and great depression is observed, indicating that the attack is an alarming one. When the disease is the result of exposure, the febrile symptoms are more often observed. One of the most noticeable changes is in the quantity of urine secreted, it becoming very scanty, high colored, and in the severer cases bloody.

The bowels are usually constipated. The animal moves about slowly and awkwardly; in getting up and lying down he appears stiff in the loins.

Pressure over the region of the kidneys will cause the animal to shrink from the touch.

Attacks running a fatal course usually terminate in convulsions and profound stupor, due to the retained poisonous constituents of the urine.

Prognosis. — The causes largely influence the results in nephritis. In simple uncomplicated attacks the chances are in favor of recovery. When a calculi exists in the kidney, or has attempted a transit from it into the bladder, and becoming lodged in the passage obstructs it, thereby destroying the active function of that important organ, the case is critical. The same or greater danger exists if nephritis results from an abscess in the kidney. When convulsions occur the chances of recovery are very slight.

Treatment. — While the kidneys are unable to perform their functions other organs must be stimulated to increased activity; the bowels must be largely depended upon. Unless grave symptoms such as convulsions or profound stupor exist, active purgatives are never admissible. It will be noted when they are employed that the action of the kidneys is always lessened by their use. It is necessary therefore to so judiciously apply treatment, that while one function is stimulated the other suffers no impairment. By administering cathartics in divided doses this end is secured.

The following powders should be given.—

$$\text{R} \quad \text{Jalapin} \qquad \text{gr.x}$$
$$\text{Sodæ Bicarb.} \quad \text{gr.xx}$$

Ft. Chart. No. xx. Sig. Dose one every two hours or until diarrhœa results.

To keep the skin active and secure its aid in elimination, the animal should be kept in a warm room and uniform temperature be maintained. Hot applications over the loins by means of the "hop bag" will be found grateful to the animal. Milk is the better diet and may be given freely, as it has an especial action on the kidneys. Beef tea, gruels, mutton broth, and scraped beef may also be given to nourish and sustain. During convalescence the tincture of the chloride of iron in fifteen drop doses is indicated.

ACUTE CYSTITIS.

Cystitis or acute inflammation of the bladder, is a disease uncommon among dogs.

Probably the affection more commonly has its origin in traumatic injuries, such as blows, kicks, and crushes. Chemical irritants, among them cantharides and turpentine injudiciously used, may induce the disease. Among other influences may be numbered chills, calculi, and long retained urine which decomposing becomes exceedingly irritating. Dogs shipped long distances and not removed from their boxes for several days, are liable to contract cystitis. An acute inflammation of the mucous membrane of the kidney may be transmitted along the passage and the bladder be involved. The disease may occur spontaneously, no obvious cause being apparent.

Symptoms. — Acute cystitis usually manifests itself suddenly. The general disturbances are in some cases marked; in others the local symptoms are the more prominent. Fever, thirst, pain, constipation, and often vomiting are not unusual.

The frequent efforts to void the urine are the first to attract attention. The quantity passed at each attempt is very small, often but a few drops, and at times the attempts are ineffectual and suppression is complete. The urine voided may be in appearance nearly normal, but more often it is thick, dark reddish in color, and blood is intermingled. The pain and distress is often great as shown by the outcries and restless movements. When standing the animal's back is arched, the abdomen somewhat hard, possibly swollen and tender over the region of the bladder.

When the attack is due to cantharides, the external genitals may be affected by the spread of the irritation.

Prognosis. — The disease usually runs a rapid course, and if uncomplicated tends to recovery. Irritation by cantharides soon subsides.

When the inflammation is due to calculi, the course will necessarily be protracted unless the cause is removed. Sometimes acute cystitis may eventuate in the chronic form of the disease.

Treatment. — If the attack is of moderate severity, reducing the diet to milk, applying hot cloths to the loins, and keeping the animal in a warm room may suffice. If the pain is severe one teaspoonful of paregoric, or fifteen drops of laudanum should be given and repeated if needed. Warm injections thrown into the bowels afford some relief.

Warm loin baths are advised if the symptoms are severe; they should be prolonged from one half an hour to two hours, their usefulness depending upon their duration.

Some authors have advised cold applications instead of warm; while the former might act well in some cases, they would be unwise in others and therefore should be cautiously used if at all.

If no water is passed, and the position and actions of the animal indicate great distress, then a surgeon should be called in to pass the catheter. After its use, sweet spirits of nitre in half teaspoonful doses should be added to the opiate.

Rectal suppositories of opium and belladonna can be used instead of giving drugs by the mouth, if vomiting occurs. These may be made as follows. —

> R   Pulv. Opii   gr.xv
>   Ext. Bellad.   gr.ij
>   Ol. Theobromæ   q. s.

Ft. Suppositories No. xij.   Sig. Use one from three to six hours as needed to subdue pain.

To introduce these properly the finger should be oiled, and by it the suppository should then be pushed well up into the rectum.

CHRONIC CYSTITIS.

Chronic inflammation of the bladder may follow the acute, and may be a consequence of irritation from calculi, from repeated excesses, and exposure to colds. Luxuriously treated house pets and old dogs are more prone to the disease.

Any impediment to the flow of urine, owing to stricture of the external passage from enlarged prostate or other causes, may give rise to obstinate cystitis. In the same manner in paralysis the disorder may be developed.

Symptoms. — The affection in the majority of cases comes on slowly. The increased frequency in passing the urine soon becomes noticeable, the quantity voided being small but nearly normal in appearance. As the disease progresses the symptoms increase in severity, the efforts to pass the water are frequent, attended with pain, and but a few drops are expelled at each attempt.

The urine now changes in its appearance; at first it is cloudy, then a mixture of mucus and pus is added, frequently tinged with blood. As the disease advances the quantity of mucus becomes great, the urine voided assumes a brownish hue, and a very offensive odor.

The movements of the animal are stiff and the gait described by some as " straddling."

The digestive organs suffer materially in chronic cystitis, strength diminishes, emaciation is progressive and the animal slowly wears away.

Under proper treatment improvement might possibly result, the symptoms gradually growing less severe, the strength returning, the urine becoming clearer and retained a longer period. Relapses will however more than likely be experienced, the mucous membrane of the bladder ulcerating, disease of the kidneys ensue and death follow.

Treatment. — As in all other morbid conditions the cause should first be removed if possible. Paralysis calls for especial treatment elsewhere advised. If calculi or other incurable conditions exist, the treatment should be directed to the palliation of symptoms.

Pain is to be overcome by the same means as recommended in acute cystitis. The diet should be largely milk and lime water, equal parts; the latter deserves especial mention as a curative agent.

The preparations of iron advised by some authors, evidently under the impression they act as astringents on the mucous membrane of the bladder, are not indicated for that purpose in chronic cystitis, for they are not absorbed as astringents, neither do they come in contact with the interior of the bladder.

One of the most active agents to be depended upon is tannin, which is thrown off by the kidneys and reaches the bladder. It may at first be given in two grain doses and gradually increased to five grains three times daily.

When the strength fails and a tonic is indicated, two grains of quinine or some form of cinchona three times a day, is advised for the peculiar and beneficial influence on the mucous membrane of the kidney and bladder. If milk seems insufficient to sustain the animal a more nourishing diet may be allowed, but it must be unstimulating.

CYSTIC CALCULI.

Stone in the bladder or cystic calculi, is occasionally met with in dogs, more common in advanced age.

Symptoms. — Irritability of the bladder, with frequent efforts to void the urine. Considerable pain exists and occasionally blood is passed; the external urinary organs are at times slightly inflamed. The symptoms may be present for a long time, not sufficiently evident to clearly indicate the exact cause which induces them. After a time a severe attack of pain comes on, the urine is retained or passes only in drops; a diagnosis then made suggests treatment, the result of which may relieve the animal somewhat, but only for a time and death finally conquers.

Treatment. — The symptoms must indicate the line of treatment to be followed. A cure is out of the question.

RENAL CALCULI.

Stone in the kidney, to which the above name is given, is occasionally found in canine practice. This affection is exceedingly difficult to diagnose unless the symptoms assume an unusual prominence.

The indications are to a certain extent those of inflammation of the kidneys; the manner of movement, the straddling gait, the arched back, and the tenderness over the loins are present as in that disease.

Pain is a prominent symptom, the urine is scanty, the efforts to void it are constant, and only a few drops result from each attempt. Hæmaturia is commonly associated. At times the suffering is intense. Fever may be present, the appetite becomes lessened, and emaciation follows.

The affection is necessarily fatal when the stone is too large to pass into the bladder through the passage called the ureter. At times when an animal has a stone in the kidney even of large size, he may suffer severely for days from the irritation it causes, then seemingly recover and be well for a long time. But other attacks occur; the stone continues to increase in size filling up the kidney, causing that organ to waste away, or an abscess to result.

The treatment is symptomatic; during an attack of pain opiates are demanded as in colic; warm baths and occasionally hot injections aid in lessening the suffering. The food should be milk as recommended in hæmaturia.

RETENTION OF URINE.

The term retention in this connection signifies a want of power to pass the urine from the bladder. It is to be understood there is urine to pass, and the condition must not be confounded with suppression in which none is passed because none is secreted.

Causation.—Retention of the urine may arise from causes functional or organic. Among the former are included paralysis or want of power in the muscular coat of the bladder, and spasmodic stricture of the urethra or canal from the bladder by which the urine passes off.

The organic causes include obstruction to the canal by contraction termed permanent stricture; stoppage of the tube, the same being blocked up with small calculi coming out from the bladder; obstruction of the tube by organic diseases as in enlarged prostate; inflammation and swelling of the mucous membranes of urethra. Possibly blood-clots may form in hæmaturia and thereby obstruct the passage. Cases have been reported where worms have lodged in the canal and closed it.

The loss of power in the muscular coat of the bladder may be due to paralysis proper, or may be induced by distention following the confinement of dogs excessively neat in their habits.

Spasmodic stricture may be caused by exposure to cold and damp, by certain drugs taken into the stomach as cantharides, or the same may be absorbed from blisters.

Temporary stricture sometimes occurs in stud dogs the result of undue sexual excitement.

Symptoms.—Restlessness and continuous pain, with constant and ineffectual efforts to urinate are the prominent symptoms. The animal's movements are unceasing and his gait stiff and "straddling." In getting up and lying down his actions are restrained as though painful. The abdomen is distended, and pressure over the bladder causes shrinking and distress. Unless relieved the pain grows more severe, vomiting occurs, the pulse runs high, becomes weak and feeble, and the general appearances indicate gravity. Convulsions often occur, followed by profound stupor and death.

Treatment.—If the symptoms are not extreme, one grain of opium should be given and followed by a hot loin bath. The great object is to arrest the efforts of the animal to urinate, and when they are discontinued, often the bladder will empty itself. After the opium a dose of castor oil should be administered.

If these measures are unsuccessfully employed, a surgeon should be called and the catheter used to evacuate the bladder.

Retention associated with paralysis demands the use of the catheter, and the employment of treatment elsewhere advised.

If the prostate is enlarged the iodide of potassium is indicated in three grain doses, three times daily. The food should be unirritating in character, milk entering largely into the diet.

HÆMATURIA.

Hæmaturia or bloody urine occurs at times in certain diseases of dogs. The seat of the disease giving rise to this affection may be in the kidneys or bladder, and very rarely in the passage from the latter.

Hemorrhage from the kidney may be caused by the irritation of calculi in that organ, or by blows or kicks over the loins; by congestion; and it might occur in certain general, serious, diseases of the system. Hemorrhage from the bladder may be caused by an inflammation of the parts called cystitis, or by the presence of calculi.

When the blood comes from the kidney it undergoes certain changes and gives to the urine a smoky appearance. It must be remembered that this appearance is not always due to hemorrhage, as certain articles taken into the stomach give rise to the same.

Blood from the bladder has a brighter appearance and is often voided in small clots. When the urethra or external passage from the bladder is the seat of trouble, blood in small drops usually follows the discharge of urine.

In hæmaturia there is associated more or less weakness about the loins; possibly some fever may be present.

Treatment. — As bloody urine is rarely more than a symptom of disease, it will demand no especial treatment, but the cause must be sought for and the infirmity removed. If a diagnosis cannot be made and the actual seat of the disorder be determined, certain general rules of treatment should be observed.

The diet first should be simplified and only milk be allowed, and of that sufficient can be taken to sustain the strength.

All medicines having a directly stimulating effect upon the kidneys should be avoided. If the manner of the animal while passing his urine is indicative of irritation or smarting, one teaspoonful of cream of tartar should be added to his milk three or four times a day, and a teaspoonful of paregoric be given occasionally. Warm applications over the loins will be grateful adjuncts.

The animal should be confined in a warm room, his bowels kept active, and the case watched until a positive diagnosis is determined.

CHAPTER IX.

DISEASES

OF

THE GENITAL ORGANS.

BALANITIS.

The principal organ of generation in the dog is subject to an inflammation of the mucous membrane of its covering or sheath called balanitis. This affection is manifested by a purulent discharge annoying to the animals themselves and exceedingly offensive to their owners.

Causation. — Local irritations such as the accumulation of certain secretions, and constitutional disorders are the more common causes.

Symptoms. — The affection gives rise to considerable discomfort inducing the animal to frequently lick the parts. The discharge is thick and yellowish in color; the mucous membrane is reddened, slightly swollen and sensitive.

Treatment. — Perfect cleanliness is the great essential. The sheath should be drawn back as far as possible and the organ bathed four or five times daily with the following. —

> ℞ Acidi Carbolici
> Acidi Tannici aa gr.xx
> Glycerinæ   ʒi
> Aquæ   ʒiij
> Ft. Mist. Sig. Lotion.

As a rule this treatment will be sufficient; if not, a solution of the sulphate of zinc, twenty grains to one half a pint of water, may be frequently used for a few days, after which a weak solution of the acetate of lead, one drachm to a pint of water, may be substituted.

Constitutional disturbances are to be overcome; if debility is associated tonics are indicated; if the animal is over fed, laxatives, restricted diet, and exercise are the measures to employ.

PARAPHYMOSIS.

This condition is said to exist when the organ of generation in the dog has protruded and fails to return to its sheath. It becomes thereby much congested and swollen inducing severe pain.

Treatment. — If the condition is recent, cold applications will often suffice and the parts retract of themselves. If force needs to be applied, sweet oil should be used freely and gentle efforts made to assist the animal. This failing a surgeon should be called to divide the constricting part of the sheath which prevents retraction.

INFLAMMATION OF THE SCROTUM.

A form of irritation of the scrotum at times occurs in dogs, which if neglected may take on a malignant type and become cancer scroti.

Causation. — The irritation is largely confined to older dogs. Of its predisposing and exciting causes but little is known. It is conjectured that general constitutional derangements aid largely in its production. When the true cancer scroti manifests itself, it is assumed that constant local irritation has no small share in inducing the disease.

Symptoms. — The usual signs of inflammation characterize the appearance of the affection. The parts become reddened, swollen, and sensitive; little pimples or pustules appear which soon discharge and a scab is formed. These scabs when removed or thrown off leave the skin beneath inflamed and very sensitive. Ulceration rapidly follows this condition unless prevented by treatment, and the tissues take on a process of induration or hardening which very likely will be permanent. The ulceration may assume a dangerous, cancerous type, progressive and resulting in a general destruction of the parts.

Treatment. — The diet should at once be restricted on the appearance of the disease, and a free cathartic be given. Cold applications are indicated, and to a pint of ice water add the following. —

$$\text{R} \quad \begin{array}{lr} \text{Acid. Acet. dil.} & \text{ʒij} \\ \text{Tinct. Opii} & \text{ʒi} \\ \text{Liq. Plumbi Subacet.} & \text{ʒiss} \end{array}$$

Ft. Mist. Sig. Add to a pint of water and use externally.

This preparation is to be applied by thoroughly wetting in it a thin cloth folded twice, and then enveloping the whole scrotum. This cloth should be frequently dampened by the solution, and may be kept in place by a bandage passed between the legs and secured around the loins in much the same way a napkin is adjusted on a child. This treatment should be persisted in until the crusts or scabs fall off. Should the skin then become ulcerated, paint the sores lightly with carbolic acid, and afterwards apply the oxide of zinc ointment.

During the local treatment it will be wise to give Fowler's solution of arsenic, four drops twice daily with the food, which should be made up largely of bread and milk, beef tea, and other unstimulating ingredients.

MORBID GROWTHS.

Not infrequently as the result of balanitis, warts or vegetations appear on the genitals of dogs. Other causes have been assigned, but in this situation they are doubtless induced by irritating discharges. They may exist as a single excrescence or as a group of several united.

Treatment. — When the warts can be ligatured it is much safer; removal by scissors or knife usually results in a troublesome hemorrhage. A silk thread should be used, well waxed or what is still better a fine elastic cord. The tying should be done at the base of the wart around its pedicle, and the string or cord be drawn tightly. This ligature will cut through in a few days and the growth drop off. When there are several grouped together, it would be better to tie but a few at a time.

The concentrated carbolic acid without admixture and applied with a brush, is especially suited to warts. The animal can be prevented from licking the parts for a few minutes and then a thorough washing will render poisoning impossible.

PROLAPSE OF THE VAGINA.

Falling or inversion of the vagina is a consequence of general debility. It is liable to follow whelping and to accompany congestive diseases of the womb. This condition is sometimes observed during œstruation, disappearing as that period passes.

The vaginal mucous membrane protrudes through the orifice having the appearance of a red, soft, and shining body, which pressure readily returns.

With the prolapse certain other symptoms are manifested. Pain and discomfort are often considerable; difficulty in passing water exists, and at times the action of the bowels is interfered with.

Treatment. — No time should be lost in returning a prolapsed vagina, as the longer the condition exists the more difficult it will be to remedy the trouble. Previous to making the attempt the protruded parts should be thoroughly cleansed, dried and oiled. After their return, vaginal injections may be used of a solution made by adding one drachm of tannin to a quart of water.

Constipation should be guarded against, as straining induces the prolapse. The diet should be mild, unstimulating and digestible.

Tonic remedies are indicated to tone up the general system. The following may be wisely given. —

$$\text{R}\quad \begin{array}{ll} \text{Pulv. Ferri.} & \text{gr.xx} \\ \text{Ext. Nucis Vomicæ} & \text{gr.v} \\ \text{Ext. Gentianæ} & \text{3 ss} \\ \text{Ext. Conii} & \text{3 ss} \end{array}$$

Ft. Pil. No. xxx. Sig. One three times daily.

If the means advised prove insufficient after a fair trial, a surgeon should be consulted and the treatment entrusted to him.

POLYPI OF THE VAGINA.

Polypi is a word employed to signify any sort of tumor having a neck or stem and growing out from a surface.

The most frequent seat of polypi is the mucous membrane; attached to the vaginal walls they are occasionally met with. Polypi in this situation usually form deep within the passage and are attached by a narrow pedicle. Their presence is rarely suspected until they appear in the orifice of the vagina. They are then the cause of considerable discomfort, and give rise to an irritating discharge. Even when of moderate size their presence is disturbing; as they enlarge they expand the vagina and compress the neighboring organs, causing urinary derangements, and may interfere with the movements of the bowels.

Vaginal polypi are pear shaped, smooth, shining, of considerable consistency, and devoid of sensation. Their attachment by a neck will exclude all doubts as to diagnosis.

Treatment. —Their treatment requires an operation. Although their removal is generally easy it would be advisable to entrust it to a surgeon who will doubtless resort to ligation.

ACUTE METRITIS.

Acute inflammation of the uterus or womb is an exceedingly rare disease. The possibility of its occurrence warrants a brief consideration.

A traumatic origin such as a blow or a kick might cause metritis, but quite unlikely. After whelping, if the labor has been long and exhausting, or surgical measures have been employed, the liability to inflammation of the womb is increased, especially if neglect or exposure follows.

Vaginal injections if too hot or too cold deserve mention as causative agents.

Symptoms.—Fever is one of the earlier symptoms. Pain is prominent and is increased on pressure over the lower part of the bowels. An examination made by introducing the finger into the vagina shows the womb to be swollen and sensitive.

Vomiting is not an uncommon occurrence. Diarrhœa and frequent efforts to urinate, are other ordinary symptoms. The movements of the animal are stiff and painful.

The lining membrane of the womb is always involved in the inflammation, and as a result a vaginal discharge soon manifests itself. This is purulent in character, rapidly assumes an extremely offensive odor, and irritates the vaginal passage and orifice which becomes hot and swollen. The disease is one of great gravity and especially so after whelping. If death does not ensue early in consequence of the purulent changes going on, the inflammation is liable to extend from the womb to the peritoneum, in which event there can be little or no hope.

Treatment.—Opiates are indicated to control the pain. Unless vomiting occurs, laudanum in twenty drop doses at first would be the wiser preparation as it is easier to vary the quantity at will.

Warm vaginal injections may be frequently given, to which if the discharge becomes fetid, carbolic acid should be added. The proportion of acid should be about two teaspoonfuls to a quart of water. If necessary to relieve constipation, rectal injections are the better means to employ; if they prove insufficient, a dessertspoonful of castor oil should be given every three hours until operative. Perfect quiet is to be enforced.

The diet should be stimulating and concentrated. Beef tea with a raw egg should be given every two or three hours. If the pulse appears weak and flagging, brandy should be added to the nourishment.

After the inflammatory stage has passed, ten drops of the fluid extract of ergot may be given every six hours, and two grains of quinine four times a day.

AFFECTIONS OF THE VULVA.

Bitches are occasionally affected with a mucous or purulent discharge from the parts at the entrance of the vagina, independent of any deeper seated inflammation. This affection resembles balanitis in the dog.

It is to be remembered that in metritis, vaginitis, and ulcerations of the womb, a purulent discharge exists, and the vulva becomes inflamed and possibly excoriated. All doubts as to the existence of these diseases are to be removed, and when vulvitis alone exists, the same remedies are to be used as advised in balanitis.

About the vulva there is liable to appear certain morbid growths, occasioning the animal considerable annoyance and pain. If small they should be cut off or, which is better, ligatured and then cauterized to prevent their reproduction.

On the external genitals of bitches there possibly may occur ulcerations and various enlargements, the natures and causes of which it will be difficult to understand; purulent discharges, want of cleanliness and proper treatment, with a generally impoverished state of the system are the more active agents.

In a general way treatment may be said to consist of perfect cleanliness, tonic remedies, astringents, and efficient cauterization.

CHAPTER X.

—•◦•—

DISEASES OF THE EYE.

—•◆•—

OPHTHALMIA.

The conjunctiva is the mucous membrane of the eye. It lines the inner surface of the eyelids, is returned and forms a covering to the exposed part of the eyeball. This membrane is subject to inflammation to which the name ophthalmia is given.

Causation.—Many causes can be assigned for the disease under consideration. Local irritation as injuries from scratches and blows, or foreign particles becoming lodged in the eye; disorders of digestion; cold and damp; inverted eyelashes; and various derangements of the general health are among the more prominent causes.

The disease may be induced by a close, damp atmosphere, saturated with animal vapour, such as may be noticed in stables improperly ventilated and neglected.

Certain forms of ophthalmia are both capable of being produced by contact with the purulent secretion, and by exposure to floating particles of pus in tainted air. As an instance, dogs may be kept in adjoining kennels with no possibility of direct communication by actual contact, and yet the atmosphere may convey impurities and the disease be transmitted.

Symptoms.—In common or catarrhal ophthalmia the symptoms are at first an intolerance of light, and a flow of tears on exposure of the eye, followed by a thin purulent discharge, which in severe cases becomes thick and possibly contagious.

The conjunctiva changes to bright scarlet red; more or less pain is present, and considerable difficulty will be encountered in making an examination. As a rule no constitutional symptoms are associated.

In purulent ophthalmia the inflammation is violent from the first, and a thick purulent discharge soon appears, often the commencement of the disease. The conjunctiva becomes intensely red; the eyelids swell and are glued together, confining the purulent secretion, and rendering constant local applications more urgent.

In the severer forms of ophthalmia there is considerable constitutional disturbance with prostration, some fever, hot nose, constipation, and the usual signs of inflammation.

Prognosis. — In milder cases the disease is very readily controlled, and a complete recovery without injury to the sight of the eye may be contemplated. In purulent ophthalmia with severer symptoms, the affection may lead to ulceration or sloughing of the anterior coat of the eye, and inflammation of the internal parts of the eyeball.

Treatment. — Great attention should be paid to cleanliness from the first, and the discharge should be continuously washed away with warm water, and a soft sponge cut in the shape of a wedge, the sharpened edge of which may be used to enter between the lids and thoroughly cleanse all parts. In the milder cases, after carefully bathing the eye, a wash made of borax and camphor water, ten grains of the former to an ounce of the latter should be used freely, and as often as possible. Once a day the following should be applied. —

R Argent. Nit.   gr.ij
   Aquæ Distil.   ℥ i
Ft. Mist. Sig. Eye wash. Apply with a camel's-hair brush.

To prevent the edges of the lids sticking together fresh lard can be used. In severer cases the bowels of the animal should be freely moved by an active purgative, and the diet reduced to milk, broths, and gruels.

Confinement in a darkened room is essential, and constant, unremitting care may be necessary to save the deeper structures of the eye from injury.

The eye should be carefully washed out every hour, and the borax and camphor water used very freely. The nitrate of silver solution may be introduced once daily, and in addition to this treatment there should be dropped into the eye morning and night, a few drops of the following mixture. —

R Atropiæ Sulph.   gr.i
   Aquæ   ℥ ss
Ft. Mist. Sig. Poison. Use with care.

After the fever has subsided the diet should be generous, and one grain of quinine should be given four times daily, and if the animal becomes debilitated, a dessertspoonful of cod liver oil should be mixed with his food.

IRITIS.

The iris (a rainbow) has received its name from the varied color it presents. It is a thin, circular shaped, contractile curtain suspended in front of the lens, being perforated by an aperture, the pupil, for the transmission of light.

The perfection of the eye is very much increased by the action of this curtain. In a strong light the pupil contracts and shuts out the superfluous rays; in a feeble light it dilates in order to admit into the eye all the light which can be received.

The iris is liable to inflammation. While the name iritis is used to indicate a distinct affection, it is rarely that inflammation is confined to the iris alone, but parts anterior to it, and the deeper structures behind it are involved.

Causation. — The affection may be the result of injuries, over-exertion, or caused by various morbid states of the blood.

Symptoms. — The white of the eye changes to a bright pink hue. The iris at first becomes indistinct, losing its color; the pupil is contracted, and its inner edges are irregular. In the next stage, one of effusion, the surface appears in some cases rusty; in others a film closes the pupil. In very acute cases the conjunctiva becomes affected by the inflammation. Pain, restlessness, intolerance of light, and excessive flow of tears are the common associate symptoms.

Prognosis. — If the disease is of recent origin the prognosis is favorable. If it has existed for a long time and came on slowly and insidiously, the chances of recovery without impairment of sight are doubtful.

Treatment. — To allay pain, and quiet restlessness, paregoric in teaspoonful doses can be given. To subdue inflammation, the bowels should be moved freely by epsom salts, and the diet be restricted to milk, broths, raw eggs, etc.

To preserve the pupil entire, it should be kept well dilated by means of a solution of the sulphate of atropine (gr. i ad. aquæ distil. ℥i.) Of this a few drops should enter the eye morning and night.

If the animal is full blooded, strong and hardy, "Gray powder" in one grain doses may properly be given three times a day between the feedings. This mercurial should only be continued for four or five days. After the acute inflammation has subsided, the iodide of potassium may be given three times daily in three grain doses. If the animal is debilitated the syrup of the iodide of iron is advised in ten drop doses at each feeding. If the need is apparent cod liver oil may also be added to the treatment. The tonics should be persisted in until the film disappears, leaving the pupil clear.

CATARACT.

The definition of the term cataract is an opacity of the crystalline lens, or of its capsule, or of both.

To render this more intelligible it is necessary to briefly consider the anatomy of this portion of the eye. The crystalline lens is a transparent, double convex body, situated immediately behind the pupil. The function of this lens is to produce distinct perception of form and outline, and to accommodate the eye to vision at different distances.

The capsule of the lens is a transparent and highly elastic membrane which closely surrounds it.

Causation. —Cataract, especially capsular, may be caused by wounds and injuries to the lens, its capsule, or to the eye itself. It is attributed to inflammation, and has been produced artificially in animals by modifying the constitution of the blood. From this it is to be inferred that certain derangements of circulation and the general system may induce cataract.

Two forms of cataract are recognized, the soft and the hard; the latter is peculiar to advanced age.

Symptoms. —In cataract of the capsule, there will be seen behind the pupil a non-transparent body of a gray, dead white; if cataract of the lens exists, the body will be bluish-white or amber colored.

After a certain age an unaffected lens acquires a yellow color, then changes to an amber.

In hard cataract the cloudiness appears in this already discolored lens, and great impairment of vision may result before the grayness is plainly evident on a superficial examination.

Treatment. —When the characteristic changes in the eye are detected, every means to improve the general health should be resorted to. Tonics if indicated are to be given, and selected as the need is manifest.

The employment of the most judicious methods can have no other influence on the disease than to possibly retard its progress. Nothing can exert a curative influence excepting the knife of a surgeon.

AMAUROSIS.

The meaning of the term amaurosis is imperfect vision, depending on changes in the deeper structures of the eye. The name "gutta serena" is sometimes given to the same affection, suggested by the peculiar glassy appearance of the eye; which while it remains clear loses its expression and becomes staring.

The manner and staggering gait of the animal suggests dimness of sight.

Causation. — Certain forms of amaurosis may be caused by long continued illnesses and other debilitating circumstances. Others may supervene upon severe digestive and liver disturbances, and possibly where worms exist.

Amaurosis sometimes arises from wounds of the head, blows, etc. Certain poisons can induce the affection.

Symptoms. — In severe cases the sight becomes suddenly dim and the animal is soon totally blind. More often the changes are gradual, and impairment of vision progresses by slow degrees. The uncertain, hesitating gait at first attracts attention. The eye then may show but little change from the normal, but as the disease progresses the pupils become dilated, the eye clearer and expressionless. Even when only one eye is at first affected, the other eventually becomes similarly diseased through sympathy.

Treatment. — In certain conditions a sudden attack of amaurosis may occur and be recovered from.

In cases where the changes are by slow degrees, the chances are the least encouraging. Were it possible to clearly determine the cause in each instance, much might result from proper medication.

In but a very few cases can positive indications for treatment be discovered, and the only evident duty will be to improve the general health, and treat symptoms.

Intestinal troubles are to be overcome; the animal treated for worms, if a suspicion of their presence exists; the body sustained and nourished; more and better blood generated; and a long course of iron and strychnine is advised after all known causes are remedied.

The following pill is a combination of these drugs. —

R Ferri Phosphas.   ℥iss

Strychniæ   gr.ss

Ft. Pil. No. xl. Sig. Dose one twice daily.

PROTRUSION OF THE EYEBALL.

This condition fortunately is but rarely encountered. Displacement of the eyeball from its socket is usually the result of direct violence received in fighting.

Simple protrusion without laceration of the attachments of the eye can, if recent, be easily overcome. The eye should be returned as soon as possible. Bathing with warm milk and water will remove dust and all impurities which may have collected on the parts, and a little sweet oil then applied renders the reduction easier. An assistant will be needed to open as widely as possible the eyelids, and it may be necessary to draw the upper one well forward with a pair of forceps. Firm, steady, but gentle pressure should be made on the eyeball, and maintained until it returns to its socket, or until evident its reduction is doubtful. When this is apparent the outer corner of the eye should be snipped with the scissors to enlarge the opening, a method which will generally prove sufficient when reduction is again attempted.

The lids should be drawn together by a stitch after the eye is returned, and the parts constantly bathed with cold water.

LACHRYMAL APPARATUS.

The tears are produced in glands situated at upper part of the cavity of the eye, and opening by ducts upon the surface of the conjunctiva between the eyeball and upper lid. The office of the tears is to keep the parts over which they are diffused moist and polished, and to preserve their transparency.

The movements of the lids spread the tears uniformly over the eyeball; they are then conducted off through the lachrymal canals, and are finally discharged into the nasal passages. The lachrymal glands are occasionally subject to inflammations.

In some affections there is an over secretion of tears, and in others a deficiency, and a consequent dryness of the eyes. In cases of the former, the eyes should be carefully examined for foreign bodies or inverted eyelashes. If the irritability apparently results from an impoverished and debilitated state of the system, tonics should be given. When the secretion of tears is scanty and the eye dry in consequence, glycerin should be occasionally applied.

Fistula lachrymalis signifies an ulcerous opening in the lachrymal sack. It is the ordinary consequence of obstruction of the nasal duct, the symptom of which is a persistent watering of the eye. Inflammation follows this stoppage, resulting in an abscess, which bursting causes the fistulous aperture from which the name of the affection is derived.

The treatment of this condition demands the skill of an experienced surgeon.

AFFECTIONS OF THE HAW.

In the inner corner of the eye is a semilunar fold of the mucous membrane, the cavity of which is directed outwards. This is called the membrana nictitans, and resembles somewhat the third eyelid in birds.

The purpose of this structure is to protect the eye, and by it dust and irritants are swept, as it were, from the eyeball.

This membrane is subject to inflammations resulting from injuries, irritants, and possibly constitutional diseases. During the existence of an inflammation the part becomes red and swollen, partially covering the eyeball obstructing the sight, and giving rise to a profuse flow of tears. Considerable pain often attends the affection, which if persistent may result in a permanent enlargement of the membrane, and consequent unsightly deformity, as well as becoming a source of inconvenience and annoyance to the animal.

Treatment. — The acute inflammation should be controlled by frequent bathing and the use of the following.—

$$\text{R} \quad \text{Zinci Sulph.} \quad \text{gr.iij}$$
$$\text{Aquæ} \quad \text{℥ij}$$

Ft. Mist. Sig. Drop into the eye three or four times daily.

This astringent lotion or the use of borax and camphor water, ten grains of the former to one ounce of the latter, will in nearly all cases prove sufficient; if not the membrane must be snipped with the scissors.

AFFECTIONS OF THE EYELIDS.

An eczematious inflammation frequently occurs on the edge of the eye-lids, they becoming encrusted with dried secretion and sticking together.

This affection may be acute, accompanied with some pain and soreness; more commonly it is chronic and associated with general eczema, obstinate in character and attended with itching. This affection is more commonly observed in debilitated animals, suffering from digestive disorders and liver derangements. It may lead to ulceration of the eyelids and disease of the roots of the hair.

The health should be restored, and local cleanliness insisted upon. Once or twice daily around the edges of the lids should be applied the following ointment.—

R Ungt. Hydrarg. Nit.   ʒi

Ungt. Simplicis   ʒij

Ft. Ungt. Sig. Apply with care, allowing none to enter the eye.

Parasites sometimes lodge about the roots of the eyelashes, and produce an obstinate itching which eventually results in an eczematious condition. The mercurial ointment advised above will destroy the insects and overcome the irritation.

CHAPTER XI.

DISEASES OF THE EAR.

CANKER.

The term canker is certainly a convenient one, as the older authors found it quite sufficient to designate a variety of affections of the ear, distinct in character, occurring in different locations, and arising from a variety of causes. Some recent writers have attempted a different classification, but their selections are equally objectionable, and it seems expedient to still recognize the term canker sanctioned by common acceptation.

The disease is really an inflammation of the lining membrane of the auditory canal or passage, which extends from the large cavity of the external ear inward to the drum membrane.

Causation. — Among the many causes which may be assigned as inducing canker are cold, digestive disorders, improper food, over feeding with insufficient exercise, the accumulation of the natural secretion in the ear, and possibly, blows.

It may accompany skin diseases, or it may be a sequel of any exhausting illness. During a long run and when overheated, dogs will frequently plunge in stagnant pools, and dirt and mud will lodge in the outer ear and give rise to the affection.

Eruptions sometimes appear on dogs suffering from a prolonged and severe attack of distemper. The passages to the ears are often invaded by these pustules, and canker results.

Symptoms. — Unless the ears of dogs are frequently and carefully examined, the affection is likely to have existed for some days before attention is attracted. The symptoms observed in an animal suffering from canker are his restlessness, frequent scratching of his ear, and violent shaking of his head. On examination the skin in the outermost part of the ear passage is found red, hot, and some swollen.

The irritation may be confined to the parts forming the external cavity, a condition designated by some as "external canker."

The tendency of the disease is to invade the innermost portions of the tube, becoming then the so called "internal canker."

The disease may have its origin either without or deep within the passage. In the latter little or no redness or swelling will be visible, but an offensive odor and a discharge is characteristic evidence.

The inflammation if uncontrolled results in an ulceration, recovery from which is often tedious and delayed.

Treatment. —It is difficult for the writer to appreciate why the older authors gave to canker that importance with which they invested it in their works on canine diseases.

No less amazing is the treatment they employed, which possibly is a solution of the enigma, for remedies such as many advised, could only have aggravated the disease and retarded recovery.

The first law to be religiously complied with is perfect cleanliness; very many cases in their early stages will yield to that treatment alone.

This fact must be remembered in the selection of medicines to be applied within the ear, no more delicate organ exists in the body, and none are more easily injured by wanton applications. An eminent aurist once thus advised a patient, who was addicted to using a small ear spoon to clear the passage. "Never put anything into your ear but your elbow," advice which if followed, would materially lessen suffering and save many their hearing unimpaired.

To insure cleanliness the outer ear should be frequently sponged with soap of an unirritating nature, and warm water. If the passage has been invaded by the inflammation, water must be gently injected, sufficient in quantity to remove all discharge and everything foreign within. It matters little the syringe used, provided it be large and easily managed; those small glass affairs are an abomination. The ear should be syringed several times daily and after the operation the following used.—

> ℞   Acid. Carbolici.   ℨss
> Glycerinæ   ℥ss
> Aquæ   ℥ijss
> Ft. Mist.   Sig. Drop a little in the ear.

Neither this nor any other application should be used cold, but it should be of the same temperature as the body.

In syringing the water should be warm; a good method to estimate properly how hot it can be comfortably borne by the animal, is for the operator to inject some against his own cheek, which is far more sensitive than his hand.

If frequent syringing does not relieve the pain and itching, it will be well to use the following—

R Bromo-Chlorali

    Tr. Opii   aa   ℥i

    Aquæ   ℥vi

Ft. Mist.   Sig. Drop in the ear.

This preparation can be alternated with the first one advised, using one, then two or three hours later the other.

When there is much discharge the powdered boracic acid dropped in the ear is productive of good results.

If canker is confined to the outer cavity and flap of the ear, accessible to direct application, stronger remedies can be safely used. If the parts are simply red and inflamed, cleanliness and applications of the following will suffice.—

       R Acid Carbolici   ℥i

          Glycerinæ   ℥i

Ft. Mist.   Sig. Paint the affected part several times daily after sponging.

If an ulceration exists, it will be well to cauterize with the stick nitrate of silver or concentrated carbolic acid, after which the paint can be applied.

While recognizing the fact that in a majority of cases of canker there is some associate general affection which demands treatment, the need to follow the stereotyped rule of first giving a cathartic is not appreciated by the writer, unless the conditions are favorable and there is good and sufficient reason for it.

If the animal is overfed and plethoric, or if some fever attends, then certainly a purgative is demanded; but if on the other hand he is debilitated by a long illness, or some rapidly exhausting disease, it were much better withheld.

In debility, iron, quinine, or cod liver oil with a generous and nutritious diet are indicated.

Whatever disease may be associated, its treatment is not influenced by the presence of canker, but constitutional remedies are to be administered as the need is manifested.

OTITIS, MEDIA OR INTERNA.

It is to be understood that the membrana tympani, or drum membrane of the ear, is the division between the so-called external and middle ear. In treating of the affection which we were obliged to call canker, owing to the difficulty and inconvenience of displacing it, inflammations of the external ear were alone considered. There remains affections of the tympanum and internal ear to be described.

Otitis is strictly the proper name to apply to inflammations of the ear in general; to this may be added terms to distinctly designate the particular locality and structure affected.

In a general way the ear may be divided thus: external, middle or internal. Still other subdivisions might be made, but they are needless in this work and would simply tend to confuse the reader. The middle ear is the space internal to the drum membrane, and the internal ear is beyond that and comprises the deeper structures. Properly inflammations of the ear should be denominated thus: otitis externa, media or interna.

As the first has already been considered as canker, there now remains to be described otitis, media or interna. In this disease the cavity of the ear becomes inflamed, involving the drum and generally the external ear, and invading the cells and inner structures. The inflammation may be confined to the mucous membrane alone, or with it the membrane covering the bone and even the bone itself may become affected.

Causation.—The usual causes are exposure to cold, sudden chills such as may result from plunging into the water and remaining too long, violent injuries and blows in the region of the ear, and foreign bodies in the external passage. Powerful chemical liquids which have been poured into the ear, accidentally or for the purpose of treatment, may so irritate the drum that deep seated inflammation will result.

The disease may have its origin in the middle ear or it may follow the profuse purulent discharge of a severe attack of canker, which finally ulcerates through the drum and gains entrance to the cavity within.

Certain constitutional conditions and disturbances furnish predisposing causes for this affection. When the blood is impoverished and the system debilitated by long tedious illnesses, otitis is more prone to occur.

Symptoms.—The disease almost always commences suddenly, attacking one ear, but never both at the same time. Pain which marks the invasion of the inflammation is intense and rapidly increasing in severity, eventually becoming so torturing as to cause the animal to shake his head violently, rub his ear along the ground, and utter incessant, sharp, shrill, ear piercing cries. He seems distracted and cannot be tranquillized.

These violent local symptoms are attended with disturbances of the general system, fever, hard and rapid pulse, great thirst, and entire loss of appetite.

In some cases the animal seems delirious, so maddening is the pain which tortures him. Great tenderness exists on the affected side, and an examination will be made with difficulty. The passage to the ear will be found more or less reddened, swollen, dry, hot, and sensitive.

These painful symptoms may last five or six days; before the expiration of this time, death may in rare instances take place from convulsions, or possibly an extension of the disease to the brain.

If the animal lives, matter forms within the middle ear, and at last the membrana tympani ulcerates, or ruptures and pus streaked with blood is discharged by the external ear. This discharge is odorless at first but in three or four days becomes highly offensive.

After an opening is made the more violent symptoms subside. In some cases the discharge persists for a time and then disappears; the hole made in the membrane closes, and a cure is affected with but little loss of hearing. More commonly the small bones of the ear are discharged with pieces of other bones which have necrosed or been destroyed, and a cure takes place with almost a complete loss of hearing. In other cases the brain may become affected by the extension of the disease and death result.

In rare instances instead of the matter discharging through the external ear, it may so affect the deeper cells and structure that a swelling back of the ear appears; an abscess forms which eventually opens and thus allows the pus to escape.

During the progress of otitis the general system suffers severely, debility becomes marked, emaciation is progressive, the animal is slowly wearing out.

Treatment. — The intense severity of the pain demands the use of opiates, and fifteen drops of laudanum should be given every two or three hours as needed. Hot applications if they can be made to the side of the head, may aid slightly in relieving the suffering. During the inflammatory stage, the bowels should be kept open by two or three teaspoonfuls of epsom salts given as needed. It is to be remembered the disease occurs more often in debilitated animals, and the effort to nourish and sustain should be early commenced.

If nourishment is not taken voluntarily, the sufferer should be forced to swallow concentrated beef tea, broths, milk, and raw eggs. Two grains of quinine may wisely be given three times a day. After the discharge appears, absolute cleanliness should be enforced. The ear should be gently syringed every two or three hours, and if much odor exists, the drops recommended in canker should be used after each injection. As the odor disappears, and if the discharge remains profuse, after syringing the ear with warm water, an astringent injection of alum and water, one half a drachm of the former to an ounce of the latter may be used.

If pieces of bone appear in the discharge, sulphate of copper five grains to an ounce of water may be injected instead of the alum.

After a time as a substitute for the medicated injections, boracic acid may be freely dusted into the ear after a thorough syringing with warm water.

As the appetite returns the most nourishing diet should be allowed, consisting largely of raw beef. Cod liver oil in dessertspoonful doses should be given three times daily, and instead of quinine, the citrate of iron and ammonia may be given in four grain doses with the oil.

By improving the tone of the blood, nourishing the body, and sustaining the strength of the animal, a cure of the local affection may be accomplished.

POLYPUS.

The term polypus includes two forms of diseased growths. The first is extremely rare; it is nearly colorless, fleshy in appearance, and attached by a stalk to the middle of the external ear. Its surface is smooth; it is devoid of sensibility, and often unattended by any discharge. This form of polypus is the result of inflammatory changes which usually subside before the growth appears.

The second form is not uncommon; it is the immediate consequence of inflammation and is invariably attended by a discharge. By some it is called the bleeding polypus or hæmatoid; it varies in size and may become sufficiently large to appear at the outer opening of the ear. Its surface is rough and glistening, and deep red in color. It is almost gelatinous in substance, is exceedingly sensitive, and bleeds easily. This form of growth has usually a deeper origin near the drum of the ear, and is more commonly a consequence of otitis, media or interna.

Symptoms. — Constitutional symptoms are present in certain cases of polypus; in others none appear. If they are large growths and attended with pain, the animal by his manner indicates the seat of the disease as in canker. The local symptoms are a profuse, offensive discharge, tinged with blood.

If a dog has experienced an attack of otitis, media or interna, and four or five weeks later, notwithstanding treatment, a discharge persists, purulent, bloody, and of very offensive odor, it is strongly indicative of the existence of a polypus of the second form. An examination with a speculum renders a diagnosis easy, even when the polypus is forming and very small.

Treatment. —Efforts should be made to improve the general health, and relieve the local condition which first induced the formation of the poly-pus. Perfect cleanliness by injections, the use of powdered alum, or boracic acid dropped into the ear, may cause the growth to waste and be-come detached. If this treatment is unsuccessful a skilled aurist should be consulted.

DEAFNESS.

Impairment or loss of hearing power is more commonly the result of organic changes. Very rarely deafness depends upon certain constitu-tional disturbances, there being an entire absence of all symptoms indicative of inflammation.

Certain drugs such as quinine, taken in very large doses, and persisted in for a long time, have been known to induce deafness. In young dogs this is generally transitory, but in older the impairment may be permanent.

The more common cause is otitis, media or interna, or some organic change in the auditory nerve, possibly induced by blows on the head, or sympathetic with disease of the brain.

In very rare instances a catarrhal inflammation of the middle ear may be excited and subside without the occurrence of purulent changes, leav-ing a chronic irritation which eventually causes a thickening of the drum membrane, and impaired hearing power in the affected ear.

Accumulations in the external passage may cause deafness, at first me-chanically, and ultimately by pressure inducing disease, and permanent injury of the parts acted upon. Some puppies are imperfect at whelp, the sense of hearing being entirely absent.

Treatment. —Excepting in cases where the deafness is due to accumu-lations in the external ear, the results of treatment will prove negative. If congenital, positively nothing can be done. If the drum membrane has been perforated, a judicious treatment of otitis will favor a closing of the opening, and the hearing power will in a measure be restored.

The deafness of old dogs admits of little or no improvement. Catarrhal inflammation of the middle ear generally results in a chronic disease very difficult to cure, and for which there is no treatment which will promise much. When organic nerve deafness exists, the condition is due to paral-ysis and should be treated as such. Where the cause is obscure or due to evident constitutional disturbances, efforts should be made to tone up the system, thereby improving the general health. Where deafness is sus-pected the ears should be well syringed and then carefully examined.

AFFECTIONS OF THE FLAP OF THE EAR.

Abscesses occasionally form in the flap of the ear. They may occur without any apparent cause but are more commonly the result of blows or tugging at the ear. The constant shaking of the head induced by canker may be sufficient to give rise to abscesses.

Their contents are usually watery in character and they have the peculiar baggy appearance of large blisters. An opening should be made in the lowest part of the abscesses; this must be large and free, and kept open until the sack has entirely closed. This treatment and perfect cleanliness will ordinarily be quite sufficient. If matter forms and a running sore results, the oxide of zinc ointment will readily heal it.

At times an eczema of the flap occurs independently of a general manifestation of the disease. This condition if not induced by constant scratching is almost always aggravated by it. The treatment of this affection is identical with that of mange.

A dry, scaly condition of the external passage of the ear involving the inner portion of the flap may exist, and often follows acute inflammations of that organ. The affection annoys the animal exceedingly, induces constant scratching. In the treatment the following will be found efficacious.—

> R Zinci Sulphocarb. gr. vi
>   Aquæ   ℥ i
> Ft. Mist. Sig. Drop in the ear three or four times daily.

At night it will be well to freely apply this ointment to the external parts affected.—

> R Ungt. Hydrarg. ℥ i
>   Ungt. Simp.   ℥ iij
> Ft. Ungt. Sig. External use.

Constitutional derangements should be sought for, and if found, are to be treated as the need is manifest.

Othæmatoma or blood tumor of the ear is a form of disease said to affect dogs, an account of its occurrence in whom, has been given by Mr. S. Ogier Ward. Wilde states that he observed the trouble in a valuable pointer.

The cat is rather more liable to the affection. One case has been described where the entire auricle was swollen out, the affected organ forming a long pointed tumor. The effusion was gradually absorbed, thickened, and shrivelled. In this instance the presence of a cutaneous trouble at the upper and back part of the auricle was noted.

CHAPTER XII.

DISEASES OF THE SKIN.

ECZEMA.

Eczema is a disease of the skin which first appears either as minute blisters, pimples, or small elevations covering pus; these three forms may more conveniently be designated the vesicular, papular, or pustular. The first is the more common form, and the vesicles are minute, transparent, and glistening; slightly elevated, and pressed together in irregular patches with little or no redness between. The fluid in the vesicles soon becomes gummy and cloudy; it may be absorbed but is more commonly discharged; adheres to the surface, dries rapidly, and forms crusts beneath which is the beginning of the disease.

The skin becomes reddened, moist, and swollen. Successive crops of vesicles form, the discharge from which is a constant source of inflammation, and the disease spreads showing different stages in different parts. The progress of the pustules are identical with that of the vesicles.

When the disease appears in the form of papules, they either change into vesicles and run the course described, or they dry into scales and crusts.

Eczema rubrum, eczema squamosum, and other terms are used to designate certain forms of the disease belonging to the subsequent stages. In the first the skin is reddened and inflamed in patches covered with shining vesicles; in the latter the eruption is dry and scaly.

Pityriasis rubra is a form of eczema rarely met with. It is characterized by a skin reddened in large patches, and covered with branny crusts or scales, which if removed the skin will be found dry and reddened, but not bleeding. This disease usually attacks the whole surface of the body, and is distinguished by its obstinacy and tendency to recur.

Eczema simplex is the term used to designate the mildest form of the disease yielding more readily to treatment. Numberless subdivisions have been made by authors who have chosen terms to designate different varieties of eczema having more or less perfectly marked stages, but they merely represent the different forms of the disease in various situations and subjected to dissimilar influences.

Eczema may be acute or chronic; the latter form is far more frequent and may attack every portion of the body, while the former may be confined to certain locations. All forms of the disease are accompanied with intolerable itching.

Causation. —Eczemas are in part primary diseases and in part symptomatic. The primary result from immediate irritation of the skin, as caused by acrid medicated applications in ointments, etc., by extremes of temperature, and by mechanical injuries affecting the skin directly. Croton oil, strong mercurial ointments, alkaline soaps, iodine, etc., when applied are frequent causes of eczema. Mechanical causes are illustrated by the irritation of the skin produced by the nails in scratching; eczema accompanying parasitic diseases is generated in this way. Prolonged, very hot baths are sufficient in some cases to induce the disease under consideration.

The symptomatic causes are not so apparent; we know that attacks of eczema occur in consequence of certain internal diseases, but their analogy has never been positively determined. Among the symptomatic causes are indigestion, injudicious feeding, want of exercise, and impoverished blood.

Eczema frequently appears as a sequel to long exhaustive illnesses, and is very often associated with liver derangements. In highly nervous temperaments there seems to exist a predisposition to eczema, and "in-breeding" has been condemned as a possible cause. That the disease is hereditary as asserted by some has never been clearly proven. That want of cleanliness might possibly induce the affection can be readily appreciated. An excess of animal food is another cause assigned, but further proof is necessary before this can be accepted as indisputable. The feeding of starchy food in excess induces disturbances of the nutritive functions, and eczema frequently results.

Worms by intestinal irritation and other influences on the system might give rise to the affection. Puppies during the period of dentition sometimes suffer eczematous attacks. Many other causes have been assigned, and many cases occur, the origin of which it is impossible to explain.

Eczema is not contagious, but in certain stages when the secretion is profuse, the disease can be communicated by actual contact of a sufficient duration. To be more explicit a healthy dog might be allowed to play with one affected with eczema, and be many times in momentary contact without acquiring the disease; but were they allowed to occupy the same sleeping-box, and the healthy skin of one remain sufficiently long in contact with the diseased skin of the other, from which there exuded a profuse secretion, through this secretion the disease could be transmitted and the previously healthy animal become eczematous.

Prognosis. —Eczema is a curable disease. An animal once attacked is liable to be again affected. In long haired dogs the disease will prove more rebellious and difficult to cure. In pityriasis rubra, certain degenerative changes occur in the skin, rendering a cure of that form of the disease exceedingly difficult.

Treatment. —While the internal treatment of eczema is positively essential in very many cases, it must not be employed indiscriminately, but only after a diagnosis of the disease associated is determined beyond a reasonable doubt. The connection with diseases of the internal organs may not be clear at first, but careful study will ultimately develop a clue sufficient to direct internal medication; until then it were wiser to depend entirely upon external remedies.

When eczema first appears especially in puppies, considerable fever is associated and a laxative is indicated. It were better to increase the activity of the bowels by divided doses rather than administer purges. Epsom salts or calcined magnesia are to be preferred; the dose of the former, two teaspoonfuls, of the latter, one teaspoonful, once or twice daily until the fever subsides, after which they are to be given cautiously if at all, as intestinal derangement and debility will result from their prolonged use.

In rare instances cases will be met with in which a feverish condition exists for a long time, the animal being full blooded and of inflammable tendency. In such cases a restricted diet and the use of laxatives is indicated; to obviate depression iron should be combined as in the following—

> R   Magnesiæ Sulphatis   ℨi
> Ferri Sulphatis   ℨi
> Acid. Sulph. Aromatici   ℨss
> Aquæ   ~   ℨviss
> Ft. Mist. Sig. One teaspoonful three times daily.

When dogs are reduced nutrition must be improved, and the most generous diet be allowed into which meat largely enters, and it may be given cooked or raw. If the loss of weight is very marked, cod liver oil in tablespoonful doses should be added to each feeding.

In cases of debility an iron tonic is indicated, and if the skin eruption has existed for sometime, Fowler's solution of arsenic may be combined as follows—

> R   Ferri et Ammon. Cit.   ℨi
> Sol. Fowleri   ℨij
> Ft. Mist. Sig. Four drops three times daily with the food.

Very many cases will be met with where this preparation of iron and arsenic, also cod liver oil, should be given at the same time. Frequently loss of appetite is a prominent symptom which demands especial treatment first, and before other internal remedies can be employed.

In such cases quinine in two grain doses four times a day should be given, and the animal urged to eat raw beef, milk, beef teas, etc., and if partaken of sparingly he should be tempted with food every few hours.

When gastric derangements exist, they should be treated as advised in indigestion.

One cause of eczema which the writer does not recall having seen mentioned, and which he desires to dwell upon, is to be found in torpor of the liver and other hepatic derangements. Cases have been observed by him which have proved obstinate, and remedy after remedy has been used without success until treatment was especially directed to the liver, and then improvement immediately followed. When a sluggish action of this organ is suspected, it would be well to discontinue other internal remedies and give the following —

> R   Mass. Hydrarg.   gr.iv
>   Pulv. Ipecac.   gr.i
>   Ext. Taraxici   ℥ss
> Ft. Pil. No. xij. Sig. Dose one three times daily.

After these pills are taken, the treatment previously employed can again be instituted. The importance of regularly enforced and sufficient exercise in plethoric, overfed animals, cannot be too strongly insisted upon.

In fact every abuse must be corrected, and every influence tending to improve the general health of the animal should be encouraged.

While the internal treatment is important, the local is no less so, and in very many cases of even greater importance. In no known disease which the canine race is heir to, has a larger number of remedies and methods been advised than for the treatment of eczema, and much harm has been done by the injudicious and indiscriminate use of irritating applications in the acute form, while much time has been lost by using the wrong application, and neglecting to employ a proper and valuable remedy.

To carefully study each case is imperative; not alone to know its causes and complications, but the stage of the eruption is of great importance, for be it remembered, remedies which are valuable in the chronic stage are not only pernicious in the acute form but do much to intensify the disease, and invite its extension over the entire body, when had judicious treatment been instituted at first, and far simpler remedies and methods been properly applied, the eruption would have remained a mere localized patch, have been of but little consequence, and immediately recovered from.

In all cases when acute eczema makes its appearance, the first step in local treatment is to remove the hair, not only over the eruption, but for a sufficient distance from it to insure the disease is seen in its entirety.

With the scissors remove the greater portion and shave closely. The parts are then to be carefully washed, remembering to use cold water, carbolic soap, a soft sponge, and to be exceedingly gentle, as an approach to hard rubbing will add to the inflammation.

In using medicinal applications it must be remembered the skin is in a very irritable state, and the treatment should necessarily be soothing. The balsam of peru is a remedy which has for years been much used in the treatment of skin diseases; in many cases it has proved efficacious, in others less active; when unsuccessful it is presumed other essentials in treatment were neglected. The writer has sufficient confidence in the remedy to recommend its use.

After the hair has been shaved off, the parts carefully bathed and allowed to dry, the balsam should be warmed and freely applied. Generally nothing more need be done until the following day, when it will be well to again use the balsam or freely apply the oxide of zinc ointment, this being especially indicated if the skin is very irritable, and the animal is inclined to scratch.

Until recovery takes place, or it is evident other local remedies must be resorted to before a cure can be accomplished, the balsam should be used continuously or alternated with the zinc ointment.

In rare instances notwithstanding these applications, the itching will still be intolerable, and the animal continue to gnaw or scratch.

If necessary camphor and morphine can be added to the zinc ointment as follows. —

$$\text{R}\quad \text{Morph. Sulph.}\qquad gr.ij$$
$$\text{Pulv. Camphoræ}\qquad \text{Ʒss}$$
$$\text{Ungt. Zinci Oxidi}\qquad \text{Ʒi}$$

Ft. Ungt. Sig. Apply not too freely.

While ointments are almost invariably serviceable, occasionally the secretion from the eruption will be excessive, and powders will be needed to check it. The powdered oxide of zinc or the subnitrate of bismuth are advised in such cases.

The writer appreciates that this advice to cut and shave the hair will not always be followed, or at least it will be done with reluctance and after other means have been sought and employed. In such cases the balsam of peru should be heated and poured on to the eruption, and rubbed in well with the finger two or three times daily, or the following can be applied —

> ℞   Bals. Peru
>   Spts. Rectificati   aa   ℥ iij
>   Zinci Oxidi   ℥ i
>   Glycerinæ   ℥ x
> Ft. Mist. Sig. Shake well. External use.

If a doubt exists as to the eruption being simple eczema, and there is a possibility that the irritation is induced by a parasite instead of the oxide of zinc, it would be well to add an ounce of sulphur to the mixture. The animal should be watched and wherever he is seen to scratch, even if no eruption is apparent the solution should be freely applied.

In the treatment of chronic eczema one must expect to occasionally encounter cases obstinate and difficult to cure; more often they will yield readily to judicious treatment. It must be remembered that a cure of eczema can seldom be effected with one means alone, but changes to others must be made as the stages and conditions vary. We can never tell with certainty how a remedy will act in every case. The skins of some dogs are more easily affected than others, and certain applications which in some allay irritation, will in others intensify it.

Notwithstanding the obstacles encountered and disappointments met, the fact that eczema is a curable disease, should encourage us to persevere.

In commencing the local treatment of chronic eczema it is to be inferred that in the progress of the disease the hair has already fallen out; if not it should be removed to admit the easy application of remedies. The crusts should be detached by the free use of oily substances; it matters but little which is selected, either fresh lard, linseed oil, cod liver oil, or vasiline. Whatever is used should be very generously applied, and in recent cases these simple remedies are often sufficient to effect a cure.

If necessary to employ other means, it would be well to gently and thoroughly wash the animal, using a strong solution of borax, or shampoo with raw eggs. After washing and drying, the zinc ointment should be freely used, and the result patiently watched. If this proves ineffectual more active remedies must be chosen. Tar has proved itself very efficacious in chronic eczema and may be used variously combined. The oil of tar and glycerin, one part to three, acts well in some cases. The thickest form of tar is generally the better, and should be combined with other agents as follows. —

> ℞   Picis Liquid.
>   Sulphur. Flor.   aa   ℥ i
>   Sapo. Viridis
>   Adepis   aa   ℥ ij
> Ft. Ungt. Sig. External use.

This ointment can be applied freely, and in cases where the hairs have not been removed, it will not stick and mat them together. A prolonged use of this preparation sometimes inflames the skin and necessitates its discontinuance for a time; in such cases it will be well to wash the animal thoroughly, and return to the oxide of zinc ointment or the use of fresh lard or vasiline. In cases proving obstinate under this treatment, it would be well to try the mixture of calomel and lime water for which the name ' black wash" is given. As a general rule tarry preparations are more appropriate when the affected parts feel stiff and rigid, and there is a tendency to the formation of fissures.

Preparations of lead are often found serviceable in the treatment of eczema; diachylon is the more common form used, and can be made into an ointment with linseed oil equal parts. Its long continued use is not advised, poisoning by absorption being possible.

The number of proprietary medicines for skin disease are countless; that each are positive cures is assured by extensive advertisements.

The writer has no disposition to discuss their value; many have virtues and others are comparatively worthless; none are infallible cures. Many of these preparations are made of the oil of tar, glycerin, and powdered sulphur; to others are added prepared chalk; still others are formed of the balsam peru and alcohol, generally in parts one to five; occasionally is found one, the basis of which is naphtha. All of these agents are active, and efficient in many cases; in as many others their use, independent of constitutional treatment, can accomplish but little.

Reviewing briefly the entire subject, we have in eczema a disease which can be cured. The fact that it has existed for a long time renders the prognosis more favorable, as in many cases the disease alone is to be combatted; not the morbid process which induced it.

In the earlier stages the soothing treatment is alone admissible. By it the intensity of the disease is lessened, even if recovery does not result.

While the eruption is localized and confined, external causes very likely induce it; on the other hand, when the eczema is more or less general, the cause is internal, and there treatment must imperatively be addressed.

In all cases the object is to correct abuses, build up the system, improve the general strength, and render the condition of the animal as near perfect as possible.

SARCOPTIC MANGE.

The sarcoptic mange is an artificial eczema occasioned in the dog by two exciting causes; one the irritation of an insect infesting the skin (Sarcoptes Canis); the other the scratching of the animal in consequence of its presence.

This insect, the Acarus, an animal parasite, is maintained by the nourishment it draws from the skin. The male acarus is much the smaller, remains in short burrows or vesicles, while the female tunnels the skin in long canals. The course of the latter is to seek a furrow on the surface, then to attach itself, and by means of its jaws penetrate the outer skin or cuticle, until it reaches the deeper and softer layers where it finds its nourishment. It continues to burrow, laying one egg after another and blocking up the passage with them. A young acarus is developed from the egg in about two weeks, and the number of eggs one insect lays is near fifty. This production goes on continuously until the female dies, which is generally in three or four months.

The young acari as soon as developed scamper over the surface, bore quickly beneath, and occasion the most intolerable itching. Around the furrows form first small pimples, which change to vesicles and pustules; from them, bloody matter exudes, which dries and produces thick, dark crusts.

The itching caused by the insects on the surface of the superficial layers of the skin incites scratching; the furrows are then opened and the acari set free. The original eruption is destroyed, but eczema displaces it, accompanied by the itching and irritation which invariably attends it; then the disease increases in extent and intensity as it progresses.

Diagnosis. — By a careful study of the eruption a diagnosis ought easily to be reached, and less difficulty will be experienced in the earlier than in the later stages. The condition of the animal and manner of attack will weigh somewhat, as eczema more commonly occurs in debilitated subjects. If other dogs in the same kennel are similarly affected, it would suggest contagion.

Vesicles in sarcoptic mange predominate over any other form of eruption at first; they occur alone and are not grouped together as in eczema; again they are pointed, not flat or rounded like those of the last named disease.

After eczema has been produced by scratching, the difficulty of diagnosis is increased, but only slightly so, for new vesicles will constantly appear on previously unaffected skin.

Treatment. — While the exciting cause of sarcoptic mange is acari, uncleanliness certainly exerts a predisposing influence, and this fact should impress itself when treatment is undertaken. When satisfied a dog has the disease under consideration, he should at once be removed from his kennel, his bedding burned, and hot lime impregnated with carbolic acid be plentifully used in every part, crack, and crevice of the building. This precautionary treatment should be renewed at intervals of every few days until the animal has recovered.

Remedies without number have been recommended for the cure of sarcoptic mange; those are the best which not only destroy the insects and their eggs and cause the secondary eruptions to disappear, but as well cure the eczema which the scratching of the animal has induced. Sulphur in some form is the sovereign remedy, united with other agents as the conditions indicate. •

In commencing treatment it is well to thoroughly wash the entire body in strong soap suds, rinsing carefully in luke warm water. The following ointment should then be used freely, covering both the affected and unaffected parts.—

> ℞ Potass. Carbonatis   ℥ i
>   Sulphur Flor.   ℥ iij
>   Glycerinæ   ℥ vi
>   Ft. Mist. Sig. External.

This preparation will immediately destroy the acari.

The balsam peru is another active agent in this disease, and may be used with the sulphur as follows. —

> ℞ Sulphur Flor.
>   Bals. Peru   aa   ℥ ss
>   Adepis   ℥ ij
>   Ft. Ungt   Sig. External.

This ointment should be applied three times a day for a week and then washed off.

Liquid storax and lard one part to two, is a favorite mixture with some. Naphthol is a very active agent and is commonly combined as follows.—

> ℞ Naphtholis   ℥ ss
>   Sapo. Viridis   ℥ ij
>   Adepis   ℥ viij
>   Ft. Ungt. Sig. External.

The entire body of the dog should be freely rubbed with this ointment once daily for two or three days, and then thoroughly washed.

In some breeds of dogs the skin is easily irritated; for them, and for house pets on which the use of ointments would be especially unpleasant to their owners, a solution may be made as follows:—flowers of sulphur two pounds; unslacked lime one pound; water two gallons; boil down to five quarts and filter, or allow it to stand until precipitated, and then pour off for use the clear fluid. This should be applied freely to the entire body and gently rubbed into the parts more affected.

Probably much less time than a week will be occupied in destroying the acari; if eczema remains, appropriate treatment should be instituted as advised elsewhere for that disease.

It should be remembered that while sarcoptic mange is purely a local affection, in nowise dependent upon constitutional disturbances; yet the annoyance caused by the itching, the loss of sleep, the effect upon the nervous system, etc., will if persistent, very probably cause derangements, which will in turn result in debility and an impairment of the general health.

FOLLICULAR MANGE.

In this form of mange the hair follicles are first the seat of the disease; the skin and subcutaneous tissues are afterwards involved by the inflammation. The affection is contagious but far less so than sarcoptic mange and its occurrence is more infrequent among dogs.

Causation. — While the disease is often communicated by immediate contact, certain conditions favor this form of parasite. Lack of cleanliness, damp kennels, improper food, and general neglect are baneful influences which predispose to the disorder. The spontaneous origin is however yet to be decided.

Symptoms. — The hair follicles first become inflamed, and the skin in proximity is infiltrated, hot, red, and tumefied; a few pimples or papules are then to be detected. The hairs from the affected follicles soon fall out, an exudation into the sacks occurs which becomes purulent, and pustules result. The pustules which are flat and run together, soon discharge their contents and scabs are formed. These harden, crack open, and bleed slightly. The eruption extends rapidly and soon the disease can be traced in its varying stages.

The animal becomes exceedingly repulsive, not only in appearance, but emits a very offensive odor.

Pain rather than itching is characteristic of the eruption. The disease may attack any portion of the body, but it usually appears first on the head.

The appetite is rarely lessened. As the disease progresses, loss of weight, debility, and impoverishment of the general system results.

Diagnosis. — When the eruption first appears a diagnosis is by no means easy, still the hot, tumefied condition of the skin, and absence of itching ought to reasonably exclude eczema and sarcoptic mange, the diseases with which follicular mange is liable to be confounded. The rapid progress of the affection, the peculiar appearance of the pustules, the crusts, and the offensive odor are diagnostic points strongly indicative. The microscope if used will remove all doubts.

Prognosis.—Follicular mange is a curable disease, but the treatment must be persevered in for a long time even after recovery appears complete, to destroy all trace of the parasitic elements, which might possibly remain on the skin and hairs and again become active. In a very severe case probably from six to eight months must elapse before a cure can be accomplished. The hair will eventually be nearly if not entirely renewed.

Treatment.—Follicular mange being solely a local affection, only remedies which act locally are needed unless symptoms of general disturbance demand constitutional treatment.

As advised in sarcoptic mange the kennel must be completely renovated. After thoroughly bathing the animal, using freely carbolic soap, the hair should be clipped and then shaved closely, exposing unaffected skin a safe distance from the eruption.

If the disease can be detected before crusts are formed, painting with the balsam peru three or four times a day would possibly be sufficient. Such good fortune in making an early diagnosis few will experience, and other treatment will be more often found essential.

When the crusts have formed it will be necessary to remove them, to favor the action of the remedies applied. The animal should be rubbed with linseed oil or common lard, a large quantity being used in the operation. The crusts will loosen in from twelve to twenty four hours. The animal should again be washed, using the domestic soft soap, and rinsing well finally.

If a house pet is under treatment and ointments are objected to, either of the following washes may be used and applied freely: sulphurous acid one part to six of water; carbolic acid two drachms, to water one pint; balsam peru one part, alcohol six parts; the lime and sulphur solution advised in sarcoptic mange is often efficacious.

Carbolic acid and Canada balsam is a preparation, which if judiciously used is most admirable; the acid enters the pustules striking at the very root of the disease; the parasite is destroyed, and the balsam lessens the irritation. If the patches of eruption are but few, the remedy may be applied in equal parts by penciling each pustule with a camel's hair brush, every two or three days or even every day. When the eruption is extensive, a few patches can be touched each day until all have been acted upon. While this treatment is being pursued, an ointment of creasote, a remedy highly endorsed by many, can be used combined with other agents as follows.—

℞	Creasoti	ℨ i
	Ungt. Hydrarg. Nit.	ℨ ij
	Sulphuris	ℨ i
	Adepis	℥ vi

Ft. Ungt. Sig. Apply quite freely to the diseased parts.

The proportion of carbolic acid to the Canada balsam might be lessened, say one to eight, and used more freely, but its efficacy would be impaired, and the stronger solution ought to be safely used without danger of absorption, as it needs merely a trace applied to the centre of each pustule.

PRURIGO.

Prurigo is a disease of the skin, manifesting itself by slightly reddish, very itchy papules of about the size of a pin head. This eruption may appear differing little or none in color from the surrounding skin, and be seen with difficulty; but to the touch it is obvious, as the finger passes over the region affected, the papules as minute elevations cause the surface to feel rough and uneven. The intense itching induces scratching; the papules are laid bare, and are succeeded by blood-red crusts of about the same size. The cause of prurigo can only be conjectured. Various speculations have been advanced; thus, certain kinds of food, worms, kidney disorders, debility, poverty of blood are said to exert predisposing influences.

The disease is chronic, and at times exceedingly obstinate. Its duration and curability are uncertain. In young animals the affection often yields very readily, but in old dogs with worn out systems, it is almost incurable.

In the treatment it is important to correct any disorders, and apply constitutional remedies as indicated. Local applications are of benefit in many cases.

When the disease attacks puppies, frequent washing, using the sapo viridis, and afterwards an ointment of sulphur and lard, is often sufficient. If the itching is uncontrollable the following may be freely applied. —

℞	Tr. Opii	ℨ i
	Creasoti	gtt.xv
	Adepis	℥ iij

Ft. Ungt, Sig. External.

In some cases the oil of tar and glycerin, one part to four, will be useful, or the tar and sulphur ointment recommended in eczema. A lotion of carbolic acid, from one to two drachms to a pint of water, or the acid incorporated with lard, from five to ten grains to the ounce, acts well in cases where the itching is intense. The internal use of iron, arsenic, and cod liver oil is advisable in cases where the disease is of long standing.

PRURITUS.

Prurigo and Pruritus are not to be confounded as being expressive of the same condition. The former is a disease always associated with papular development, while the latter is intense itching, purely a nervous manifestation, unattended with any visible changes in the skin or tissues.

Pruritus is a special form of skin irritation connected with many physiological changes. Thus, in diseases of the kidneys, jaundice, and other liver disorders, plethora due to over feeding and want of exercise, piles, worms, constipation, indigestion, and a sluggish cutaneous circulation from lack of cleanliness and proper grooming, are influences which induce pruritus. The affection is not uncommon in old dogs dependent upon the invariable changes in the skin induced by age.

In young dogs the disease is more common among those of a highly nervous temperament, possibly the result of continued inbreeding. When intense itching exists the animal must be very carefully examined to determine if possible the cause. Pruritus must not be accepted as the diagnosis until all doubt as to the existence of diseases of the skin and parasites is removed.

The condition of the animal will suggest what constitutional treatment to employ. All disorders are to be overcome and the general health improved by medicine, dieting, proper exercise, etc.

Frequent bathing or immersions in cold water when the weather permits, or the use of diluted vinegar or alcohol as lotions, are the local treatments to employ.

FAVUS.

This disease for which a variety of names is given, among them honeycomb, ringworm, is characterized by the formation of sulphur-yellow crusts. These are generally rounded and present a peculiar appearance; in their earlier stages they resemble much the eruption in ringworm, namely: thin, roundish scales pierced by a hair. These scales then form in their centre a yellow prominence which rapidly increases in size and at their bases the surface is slightly hollow; thus nearly all present an outer concave and an inner convex surface, what is called a crab's eye appearance.

The crusts throw off a peculiar musty odor, and when removed a cuplike depression is found in the skin, which may be ulcerated or have but a thin scarf-skin covering it.

In some instances the disease appears as minute vesicles, formed in circles, which dry to yellow crusts (scutula).

Favus is a disease rarely seen in the dog; it is the result of a parasite which penetrates deeply; the hairs then becoming brittle, lose their lustre and finally fall out. Itching is not a symptom, the eruption being attended with pain. Damp kennels and general neglect favor this affection as they do other parasitic diseases of a similar nature.

Treatment. — The crusts should first be soaked with oil or lard, then removed, and the hair shaved. Many local remedies have been advised; among them the oleate of copper promises especially well. The sulphuret of potassium is an agent of great value in parasitic diseases, and if the eruption is not extensive may be applied in the following compound. —

$$\text{R}\quad \text{Potassii Sulphureti} \quad \text{3i}$$
$$\text{Sapo. Viridis} \quad \text{3ss}$$
$$\text{Adepis} \quad \text{3iss}$$

Ft. Ungt. Sig. Apply three times daily.

Where the disease is extensive, the quantity of potassium should be lessened one half or one fourth as it appears judicious.

The solution of carbolic acid and Canada balsam advised in follicular mange, will prove as destructive in favus as in that disease, and if used, the same caution is to be observed. The hair will be renewed excepting where the crusts have persisted too long, and by pressure on the skin induced ulceration, which in healing has left a scar.

Ringworm.

This affection is occasionally met with in canine practice. It is due to a parasite identical with that in favus, and is the exciting element in both diseases. The same influences are active in propagating all forms of vegetable parasites or fungi; warmth and moisture favor them, and thus damp kennels, unclean bedding, and general neglect are predisposing causes.

Ringworm is very easily communicated by contact; animals may transmit the affection to man or by turn be infected. The disease manifests itself in round patches which are covered with thin scales or crusts easily detached. It spreads quite uniformly, enlarging its circle on the outer edge of which the eruption is more prominent, receding as it were from the centre. The hair becomes dry and harsh, losing its elasticity and breaks off or falls out.

Ringworm may appear in the form of small pointed vesicles, containing a clear fluid, and arranged in a circular form; or it may take the form of pale, red spots, having a small whitish scale in their centre.

Treatment. —The destruction of the parasite which excites the disease ought not be difficult. The hair should be shaved sufficiently beyond the eruption to insure all affected parts are exposed; then crusts are to be removed by soaking with oil or lard, after which the animal should be thoroughly washed in strong soap suds.

The local applications advised are numberless; among the more efficient are the black wash, tincture of iodine, and balsam of peru. A very successful method is to paint the spots well every two days for a week with the tincture of iodine, applying several coats at each operation; on the intervening days, and after the week has passed, a preparation of carbolic acid grains twenty; oxide of zinc ointment one ounce, should be used.

If the ringworm has existed a long time and the parts are infiltrated, the oil of tar should be applied after the parasite has been destroyed, and may be combined with glycerin, one part to three. Only a short time will be needed to effect a cure. After exposure of an animal to the disease, from one to two weeks is required for its development.

ERYTHEMA.

Erythema is an inflammation of the skin which appears either in the form of superficially inflamed patches, pimples, or lumps called nodules.

The accompanying symptoms are redness, some tenderness, and a slight itching. The eruption may disappear spontaneously leaving no trace, or the skin may burst, a watery discharge follow, and crusts form.

All parts of the body are liable to be attacked, but the head and extremities are more commonly affected.

Writers differ materially as to the cause of erythema; it is generally believed to be due to local influences, and to constitutional disturbances. Unclean bedding, hair long and matted, retaining the decomposing excretions from the skin, heat, cold, friction, and gastric irritation are exciting influences. Erythema is a non-contagious affection.

Treatment.—If a cause can be determined it should be combatted; cleanliness is one great essential; the skin should be kept clean by frequent bathing, after which the oxide of zinc should be applied, either in powdered form or as an ointment.

Matted hair should be removed by clipping, and appropriate treatment is to be instituted if constitutional disturbances exist.

ERYSIPELAS.

Erysipelas is an inflammation of the skin which tends to spread rapidly over large surfaces, accompanied by severe constitutional symptoms.

The disease is assumed to be blood poisoning and may be caused by contagion, or by local infection as in wounds and diseases of the skin. Other causes have been assigned by certain writers but the most eminent authorities maintain the disease is never spontaneous, but to produce it there must be some irritating matter with inflammatory properties introduced into the system, and circulated in the skin.

Erysipelas is known by redness of the skin, some swelling, heat, and severe pain. In rare instances the eruption may remain confined to the spot first affected, but it usually spreads rapidly, disappearing in the parts first attacked as it extends. The constitutional symptoms are often very severe. The disease is generally ushered in by a chill as shown by shivering; then follows pain, rapid, bounding pulse, fever, thirst, and often vomiting. As the eruption extends these symptoms become more prominent and may assume a low typhoidal type. The disease is one of great severity, and dangerous unless the eruption remains confined to the spot where it first appeared. If it can be arrested early the chances of recovery are much more favorable.

The general condition of the animal must be considered in treatment. If plethoric and of inflammable tendency a cathartic should be given on the first occurrence of the eruption. Throughout the course of the disease, supportive measures are to be employed. The diet should be nutritious and concentrated, given more generously as the disease progresses; milk, beef tea, raw eggs, and scraped raw beef are the principal articles to be depended upon. Pain should be controlled by laudanum in fifteen drop doses, repeated as the need is apparent.

In very severe cases the tincture of the chloride of iron should be given in ten drop doses, every three hours in a little water. If the animal shows signs of sinking, two teaspoonfuls of brandy should be given from two to four hours. Quinine in one or two grain doses may be added to the treatment if the pulse becomes soft and weak. Local measures are to be employed to allay the irritation, and arrest the extension of the disease. While fever exists ice cold preparations are preferable, and the affected parts should be kept constantly wet with the following solution. — Acid. Acet. Dil. ℥ij; Liq. Plumbi Subacet. ℥ij; Tr. Opii ℥i; Aquæ ad. Oi. Ft. Mist. Sig. External use only.

As the fever abates cold applications may be displaced by an ointment made of carbolic acid one drachm, the oxide of zinc ointment six ounces.

If pus forms it should be evacuated by incisions, several small ones being advised rather than a larger opening. After the operation the wounds are to be treated with flaxseed poultices as in cases of abscesses.

If the stomach becomes much disturbed and vomiting frequent, the diet must be reduced to milk and lime water, the tonics be temporarily suspended, and five grain doses of bismuth given every two hours.

CHAPTER XIII.

INTESTINAL PARASITES.

A study of the anatomical structure of worms, their physiological phenomena, and natural history is highly interesting, and productive of valuable results; still those considerations which relate directly to diseased states in consequence of their presence, will be of far greater practical importance to the reader.

Dr. Cobbold in his able and exhaustive treatise on "The Internal Parasites of our Domesticated Animals" states, "when people speak of worms in the dog, they commonly refer to round and tape worms; and in place of recognizing as they might, fully a score or more of intestinal parasites, they are content to roll the entire series into three or four species only. Thus, of the so-called lumbricoid and filariform worms, we have no less than eight or nine distinct forms, and of these the most common species is the marginated round worm."

In the recognition and treatment of disorders consequent upon worms it would seem wiser to leave the scientific study to zoologists and specialists, and to avoid confusion to reduce the classification in this work to the limit of practicability.

Probably no known disorder the dog is heir to is so destructive as intestinal worms, it being estimated by reliable authorities, that at least three fourths of the whole canine race are infected by the pest.

Among the many different species of internal parasites some are found in the liver, others in the eye; the giant strongle although rare, has been known to infest the kidneys; the wrinkled thread-worm sometimes reaches the bladder; the cruel thread-worm selects the heart, death resulting suddenly in a convulsion, or deferred for a time, during which the dog is racked by agonizing pain; in the nasal cavities and even in the blood and muscular tissues, parasites have been found.

These facts are of interest but of little practical importance, and it is by far more essential to consider those which are developed in the stomach and intestines.

ASCARIS MARGINATA.

This lumbricoid is the principal round-worm found in dogs; it resembles the common earth-worm, and varies in length from two to six inches. It is of a pale pink color, perfectly round in shape, and tapers towards each extremity.

Round-worms inhabit the small intestines, often migrating into the stomach, from which they are expelled by vomiting; but more often they seek the large intestine and pass out with the discharges. In rare instances they have been known to wander into the gall bladder and biliary ducts, giving rise to abscess of the liver; even the pancreatic duct has not escaped invasion.

That certain special conditions are required for the formation of round-worms is evident from the fact that they are more common in puppies and young dogs. As regards the nature of the conditions, little or nothing is known; mucus in abundance is supposed to be the repository of the ova or eggs from which they are propagated. The female round-worm is marvelously productive. Eschright estimated in the body of the female lumbricoid found in the intestines of man, the number of eggs to be sixty-four millions. These are discharged with the feces and retain their vitality for many months. It is conjectured that ova are introduced into the intestinal canal through the medium of the drinking water and food.

Symptoms denoting the existence of these or other worms within the intestinal canal are in some instances obscure, in others prominent. At times but little derangement is noted; again profound disturbance results.

The morbid effects of worms in dogs cannot be exaggerated; that they may give rise to convulsions, chorea, paralysis, and other affections of the nervous system is very generally conceded.

The more common symptoms denoting worms are great abdominal distension or bloating, vastly disproportionate to the amount of food taken, indigestion, colic pains, diarrhœa, vomiting, emaciation, voracious appetite, nose hot and dry, cough, offensive odor to the breath, disturbed, dreamy sleep, and a rough, dry, harsh coat.

Frequently puppies and more rarely old dogs partially lose the power of their hind legs, and rapidly recover as soon as a discharge of worms occurs. Paralysis induced by this cause is functional and rarely if ever complete.

Treatment.—For the expulsion of the round-worm numberless drugs have been advised. It must be remembered that natures and conditions vary; remedies which in some cases are effectual in others are inactive; for this reason, when there is presumptive evidence of the existence of worms, absolute dependence should not be placed upon any one agent, but others should be given after a proper interval has been allowed.

Treatment for round-worms should always be premised by fasting and a purgative. Food should be denied puppies for twelve hours at least, and old dogs twenty-four hours. Castor oil is the better purgative to adminis- ter, and should be given six or eight hours before the worm medicine is employed.

Among the more reliable agents destructive of round-worms, santonine in two grain doses, is one of the most active; it is insoluble or nearly so, and passes into the small intestines and acts on them as a poison. It will be easily taken in pill form mixed with lard, and can be given two or three times a day; after the second or third day a generous dose of castor oil should be administered and the santonine discontinued.

The oil of turpentine as a vermifuge is highly esteemed, especially in tape-worm. In cases of round-worms which do not yield to other agents, this remedy could be tried. Its action is to destroy or debilitate the parasite, which losing its hold upon the bowels, is then easily discharged. When worms exist in the stomach they are by its action killed, and then digested as any other dead animal matter.

It is to be remembered that turpentine in small doses, frequently re- peated, stimulates the kidneys, increasing the secretion of urine, and often producing if long continued, painful irritation of the urinary pas- sages, amounting sometimes to complete stoppage. Large doses generally act speedily on the bowels, in which case the oil is hurried out, and no time being allowed for absorption, it is less liable to irritate the kidneys and bladder than in small and repeated doses.

When proper precautions, suggested by the physiological action of tur- pentine, are observed, the remedy is not only harmless but very efficient; fifteen drops can be given twice a day for a week without injury. If at the end of this time worms are not expelled, it would be well to give one final large dose of one teaspoonful, and if it does not operate freely in two hours then administer castor oil. The turpentine can be given in milk or beaten up with a raw egg and a little sweet oil.

Pinkroot is a drug which has been much used as a vermifuge, and is reasonably safe and reliable. In very large doses it has a narcotic effect, but this is altogether obviated by combining it with cathartics. The fluid extract of pinkroot and senna is a combination which acts admirably in many cases of worms in puppies, causing little or no gastric disturbance. The dose for them when four or five weeks old is fifteen drops; for matured dogs, from one to two teaspoonsful in a little water. In all cases this rem- edy should be given twice a day for three or four days, after which a generous dose of castor oil should be administered.

In giving medicines for the destruction of worms, certain precautions in feeding should be observed. If but one, and a large dose of a drug is to be depended upon, food should be withheld until after the bowels have moved freely. If the drug has to be given three or four days, the diet should be simple and as limited as the condition of the animal will permit, and as long an interval as possible between the feeding and the administration of the medicine be allowed.

Other drugs, among them areca nut, are to a certain extent active in the destruction of the round-worm; they are however more deadly to the tape-worm and will be considered among the remedies advised for the removal of that pest.

Considering the great fatality among puppies caused by worms, not alone of the greatest importance is the treatment after their existence is obvious, but a means preventive is equally as urgently demanded.

Charcoal is an agent easily obtained, perfectly harmless, and of value not only as a vermifuge but as a preventive. The digestive organs of very young puppies are easily disordered, but few drugs can be well borne without discomfort, or temporary derangement, and undoubtedly many are sacrificed to injudicious dosing.

Charcoal in a reasonable quantity is not only incapable of injuring even the weakest stomach, but is an admirable agent in indigestion and intestinal irritation. As these disorders are commonly induced by the presence of worms; had charcoal no other action than to merely allay irritation and promote digestion, its administration would be advisable. It has however another and decided action on intestinal worms in young puppies, causing their expulsion and exerting a preventive influence as well. In what manner this agent acts is unexplained, it may be through its known power as a disorganizer of animal and vegetable bodies, or by destroying the conditions that favor the propagation of the ova.

Of its action on worms in matured animals, the writer has not from experience a knowledge. It has long been his custom to use powdered charcoal in the treatment of very young puppies, mixing it with their food several times a week; dose immediately after weaning, from one fourth to one half a teaspoonful; for matured animals, the dose is two teaspoonsful. The granular form is advised for older dogs.

Charcoal is much less active in immediately expelling worms than many other agents, and where their presence is indicated by marked symptoms, the more powerful remedies should be used. When this urgent need is not manifest and yet the existence of worms is suspected, the administration of charcoal every day for a week or longer, can do no harm and may be productive of much benefit. As a preventive it should be given with the food two or three times a week.

TÆNIA.

The tænia or tape-worm is distinguished, as the names imply, by its ribbon like form. It is composed of numerous joints, each of which in the complete state is provided with male and female generative organs. It is really a collection of animals, having an alternate generative power. From the head are produced the segments or joints by a process of budding.

The head is provided with suckers, and in some instances with hooks, by which means it firmly fixes itself to the mucous membrane of the intestine into which it has been introduced. The segments then multiply, lengthening the chain as it were, and increase in size and sexual development. They remain joined together until fully matured, and then separate from the colony. They generate eggs which contain the embryo or germ from which other tænia are developed. If these enter the stomach of a suitable animal, their envelopes become softened or ruptured and the embryos are set free. In some way or other they leave the digestive canal and make their way to different parts of the body, meeting with conditions favorable to their development. Should these now be introduced into the intestinal canal of another animal, they would fasten themselves to the mucous membrane and again a collection or colony of tape-worms becomes developed.

To render this method of transmission clearer by illustration, the egg from the tænia of the dog if received into the body of a sheep, there finds in certain parts of the organism the conditions necessary for its development and growth, and becomes what is known as the cœnurus cerebralis, a parasite found in the sheep's brain, which if eaten by the dog becomes the tænia found in the intestine of that animal.

Another method of propagation is illustrative; segments of a tape-worm are passed by a dog so infested, or they make their escape from the bowels and locate themselves among the hairs of the coat and deposit their eggs. If the animal is unfortunate enough to be affected with lice (Trichodectes Canis), these eggs are swallowed by them. Within the bodies of the lice the eggs meet with conditions which favor the rupture of their envelopes, and the embryos escape and another transformation takes place. In biting the parts irritated by them the lice are often swallowed by the dog, and thus the germ enters and is developed into a perfect tænia in the intestinal canal which it left as an egg but a few weeks previous. Through this method of transmission an animal may continually infect himself, or by depositing the lice containing the embryos in the kennel, shaking them from his coat into the drinking water or food, they may be introduced into the bodies of other dogs, and they in turn become infected.

Other parasites besides lice act as mediums and supply the conditions favorable to transmission and propagation.

The history of the development of tape-worm in a general way suggests the manner in which they may be acquired. It is necessary that a living embryo be first introduced into the stomach. This usually happens from eating infested meat, which has not been sufficiently cooked to destroy the embryos or render them incapable of further development.

It is of interest to know that in certain countries, tænia with but rare exceptions infest all the inhabitants, among whom raw cow's flesh is looked upon as the greatest delicacy. In this instance both men and cattle furnish the conditions favorable to propagation.

Several varieties of tape-worm infest the canine race, and are found at every period of life although most rarely among puppies. As regards medical practice, the discrimination is of but little importance, the same measures of treatment being alike applicable to all.

The most common form is cucumerine (Tænia Cucumerina). This is a delicate tape-worm which measures from ten to twenty inches in length. The anterior portion of the body is like a thread, and the segments are short but lengthen towards the tail. As they ripen the divisions between the segments become more marked so that the worm presents a chain-like appearance. They move about actively in the intestinal canal, and are expelled with the feces or escape of themselves.

The largest tape-worm found in the dog is the marginated, (Tænia Marginata) which may reach ten feet in length and the germ from which it is propagated is derived from the sheep.

Among certain sporting dogs the tænia serrata is the more common, and the parasite from which it is developed is furnished by the hare and rabbit. This tape-worm varies from two to three feet in length. Many other varieties exist, a study of which would be interesting but of little practical benefit.

Symptoms.—The symptoms especially due to the presence of tape-worm are but vague and indefinite. Disturbances of digestion, nutrition, and of the nervous system occur, but in many instances are not sufficiently prominent to clearly indicate the existence of tænia.

When the appetite is voracious, and the animal seems strong, active, and apparently in good health yet poor in flesh, if tonic treatment is employed and still the animal remains emaciated, then the presence of a tape-worm should be strongly suspected, and the discharges from the bowels carefully examined. The appearance of segments among the feces will be of course conclusive evidence, and yet it must be remembered that some are so small, others so delicate, they are easily overlooked; again it is doubted if some forms of the parasites are expelled spontaneously.

In summarizing, it is advised that all facts be weighed carefully, and every effort made to positively determine the existence or non-existence of tape-worm; then if in doubt, give the animal the benefit of it and administer worm remedies.

Treatment. —A cure of tape-worm can only be complete when the head is expelled. After treatment has been employed the discharges should be carefully examined to positively determine if it has been successful. By a careful use of water the solid constituents can be washed away and the worm exposed. If the head is not found or escapes observation, the question of complete cure must remain undecided for two or three months, during which period the discharges from the intestinal canal should be occasionally examined for segments. If at the end of that time none have been detected, it may be concluded the cure is perfect, as had the head been retained the worm would have again reached maturity.

It is judicious to employ preparatory treatment before an actual cure is attempted.

The object is to empty thereby the intestinal canal, so that the worm may be detached and expelled more quickly. The animal if matured should be denied solid food for twenty-four hours, being allowed only a very little milk or broth, and a dose of castor oil ought to be administered the day previous to the giving of the vermifuge. A stronger purgative is not judicious as it is liable to disjoint the worm and the remaining attached portion will be with greater difficulty expelled.

Among the most active agents destructive of tape-worm are areca nut, turpentine, koosso, male fern, kamala, and santonine.

Areca nut, or betel nut as it is called, can be given with perfect safety. It should be obtained in the solid form and the darker colored selected as being much stronger than the lighter. It should only be reduced to a powder by grating just before required for use. The dose is one drachm, and is better given in the form of a large pill or bolus incorporated with lard.

The manner of giving turpentine and santonine has already been described in the treatment of round-worms.

Koosso operates exclusively as a poison on the tape-worm and seems equally as effectual in all varieties. The dose is two drachms and before administration should be mixed with a cup of warm water and allowed to stand for fifteen minutes.

Male fern was very much used by the ancients as a vermifuge. Its especial efficacy is in the treatment of tape-worm, upon which it acts as a poison.

The medicine may be given in the form of a powder, but the etherial extract (oil of fern) is to be preferred. The dose of the latter is from ten to fifteen drops and should be given well beaten into a raw egg. The method of some is to administer ten drops at night and repeat in the morning, and followed at the interval of an hour, by a generous dose of castor oil.

Kamala is a vermifuge strongly urged by many. The dose is one and one-half drachms, and can be given suspended in milk. Other drugs have been advised, and for each especial merits have been claimed, but a description of them is unnecessary as the most reliable agents have been selected.

If two or three hours after giving the worm-cure the bowels do not move actively, a full dose of castor oil should be administered.

It must then be appreciated that the digestive organs have suffered rather rough usage, and for a few days the diet should be easily digestible, consisting largely of milk and broths.

After considering the manner in which tænia are propagated in dogs, the means of prevention suggests themselves.

CHAPTER XIV.

DISEASES

OF

THE NERVOUS SYSTEM.

HYDROPHOBIA.

The antiquity of hydrophobia is not precisely known. Aristotle is the first author to mention this malady, but his account of it is remarkably incorrect, if the text be not corrupted. He says, "all animals that are bitten by a rabid dog are affected with the disease except man; and that the disease proves fatal to all animals but man."

Among the ancient authors Cælius Aurelianus treated all the important questions relating to hydrophobia in a most masterly manner. After him centuries passed during which, with but few exceptions, independent observations ceased, and little or nothing was added to the previous knowledge on the subject.

Investigations were renewed towards the end of the last century by Hunter and other intelligent observers, and recently a more exact knowledge of the disease has been acquired through the studies and experiments of Hertwig, Meynell, Youatt, Magendie, Blaine, Virchow, Reder and others.

Hydrophobia or more properly rabies, which prevails chiefly among animals of the canine species, (dog, wolf, fox, jackal,) is at the present time accepted as being an acute infectious disease, coming on in the form of a functional disturbance of the central nervous system, without structural changes which can be considered essential to the affection.

Examinations after death have revealed in most cases diseased appearances, but not one has been found to be present invariably, and no local changes have been discovered which could explain all the symptoms.

Causation. —The theory of a spontaneous development of rabies so frequently advanced, has been rejected as entirely unfounded by very many of the most intelligent observers, recent as well as old.

It is unnecessary to enumerate and describe all the influences which have been cited as predisposing and accidental causes; in wearisome arguments our best authorities have proved the theories untenable and unfounded, and that all the causative conditions such as seasons of the year, extremes of temperature, restraint, starvation, suppressed sexual appetite, age, sex, and race are to be regarded neither as direct nor predisposing causes.

It is now universally accepted that the poison is communicated almost invariably by means of the bite of a rabid or infected animal.

Nothing is known of the specific infecting principle of rabies. The virus is contained in the saliva and foam of the diseased animal, also in the spinal cord, blood, and salivary glands; its existence in other portions of the body is conjectured but not positively determined. Neither by chemical or microscopic analysis of the saliva of rabid dogs has this virus been detected.

It increases by internal growth, and from other poisons it is distinguished principally by this circumstance, that it remains within the vital organism for weeks and even months without producing any diseased symptoms whatever.

In what manner, or by what course the specific poison penetrates the body from the wound is unknown, neither has its action while within the system been adequately explained.

Two theories have been advanced; either the virus remains awhile concealed and inactive at the seat of the wound or point of inoculation, and only after a certain interval — at the expiration of the period of incubation — enters and circulates with the blood and other fluids of the body; or else the poison, by undergoing incessant reproduction, is constantly supplied in fresh quantities to the blood.

Virchow has compared the action of the poison to that of a ferment, fresh particles of which are constantly being conveyed into the blood from the seat of the inoculation, producing through the medium of the circulation the specific effect upon the nervous system.

Experiments have proved that the bite of an infected—though apparently healthy—dog, when inflicted during the period of incubation, or slow development of the disease, has even then the power of communicating it. Thamhayn collated nineteen cases, occurring in the human subject, in which dogs, to all appearances healthy, but which subsequently became rabid, produced by their bite hydrophobia, the result being fatal in eighteen cases.

The virus is probably capable of infection for some time after death but hardly longer than twenty-four hours. That the same is inactive when brought in contact with the unbroken mucous membrane of the digestive canal, is confirmed by the experiments of Hertwig; at one time he introduced the saliva and mucus of a rabid dog into the mouths and throats of healthy animals; again he fed dogs on food with which not only the infected saliva and mucus had been mixed, but to which also warm blood taken from rabid animals had been added, all with negative results. He placed healthy dogs in stalls where others affected with rabies had just previously been kept, so that they were brought in frequent contact with the same straw, chains, food, and drinking basins that had been used by the diseased animals, but in no instance was rabies produced; and the same may be said of placing healthy animals in the stall with dogs which had recently died of the disease.

It is an acknowledged fact, demonstrated by numerous experiments that the milk and flesh of rabid animals (dogs, sheep, cattle) may, as a rule be consumed by man and animals without any ill effect.

The theory that the poison is very rarely communicated in any other way than by the bite of a rabid animal, is fully confirmed by the most wearisome experiments. No instance is as yet known where the virus has been transferred by intermediate vehicles. An ingenious theory has been suggested, that if minute particles of the poison suffice to propagate infection, possibly certain parasites, fleas, lice, which are nourished by the blood of the dog, and which the more often infest them, may transfer the poison by means of their blood-drawing apparatus, and thus produce inoculation. It is inferred that an analogy exists with other infectious diseases, as for instance small-pox which has been unquestionably transported by flies, and the same is true in cases of malignant pustule. In the latter contagious and very fatal malady, it is believed that flies which have alighted on the ulcers of diseased animals convey the virus, and infect other animals and human beings.

Little or nothing is known as to a predisposition of rabies, it being estimated that less than two-thirds of the animals inoculated or bitten are infected by the disease. According to the experiments of some, a rich feeding of the dogs that had been inoculated, favored the outbreak of the disease, while on the other hand it was retarded by poor nourishment. Many dogs have resisted an infection; in three years Hertwig made nine attempts to inoculate his famous poodle. Some have been known to have been bitten four times by rabid animals and still remain unaffected, as in the case of a dog in the veterinary school at Lyons.

Symptoms.—The wound from the bite of a rabid animal heals very readily, being but rarely accompanied by inflammation.

The period of incubation, or development of the disease, is variously estimated by different authors; in the majority of cases it lasts from three to five weeks. The two extremes in a dog which have been reported are the shortest one week, the longest eight months.

Rabies assumes two forms; the violent, and the dumb or sullen. Hertwig has said that two cases rarely correspond. From this it can be appreciated that an accurate description of a disease presenting so many varieties and depending on so many influences, such as age, temperament, condition, etc., is no easy task.

The violent or furious has been divided by some authors into three stages; the melancholic, irritative, and paralytic.

At the outset of the disease the animal manifests a changed manner, becoming irritable, sullen, and nervous; his disposition towards those around him is capricious; to his master he is often even more friendly and confiding, but from others, of whom he seems suspicious, he shrinks, and is easily enraged by their interference. He is disposed to shun companionship, slinking away in dark and obscure places, and when called, comes reluctantly, his manner crouching and frightened. In this stage the eyes are slightly reddened and wear a changed expression as has been described, a vacant, far-away, listless look; often they change and in consequence of the wrinkles in the forehead, the gaze becomes sullen and ferocious. It is difficult to fix his attention for more than a moment, when his eyes will close in a sleepy manner, and remain shut for several seconds.

According to the most experienced observers, a perverted appetite is one of the most constant morbid symptoms. Food is the more often rejected or eaten sparingly, and at times it is taken into the mouth and dropped again. On the other hand, all sorts of indigestible substances — sticks, straw, rags, earth, hair, dung and the like are swallowed.

The desire to chew something is irresistible, due possibly to a peculiar feeling in the jaws excited by an irritation of the nerves. If allowed in the house he will bite and worry chair-legs, carpets, boots, etc., in fact anything he can fix his jaws upon. A tendency to lap cold objects such as stones, iron, and his own urine is observed. In some cases the sexual desire seems stimulated, as indicated by the disposition to lick the genitals of other dogs. A mucous discharge appears at the nose, and the secretions in the mouth and throat become thick and ropy, which the animal will occasionally make efforts to expel, and possibly attempt to vomit. This trouble about the throat has led to a mistaken diagnosis of a bone being lodged there.

The duration of this stage denominated the melancholic, may be but a few hours, and rarely over two or three days. It must be remembered that not all the symptoms are uniformly present; in some cases they are few and insignificant, possibly overlooked, or if detected sufficient importance is not attached to them; a fact which renders this period in the disease the most dangerous to man.

The so-called irritative or maniacal stage has more defined and characteristic symptoms which appear spasmodically, with intervals between the attacks, during which the animal is far less violent. Among the most important symptoms are a changed behavior, an irresistable disposition to bite, a peculiar bark, and repeated violent efforts to break away and stray about. Of this latter symptom, Hertwig has observed that the desertion of his home by a previously faithful dog indicates the existence of a high degree of mental disturbance. The propensity to stray off, especially after having been excited or corrected is then often ascribed to fear, obstinacy, disobedience, or unsatisfied sexual desire, and the animal is all the more dangerous in this state if he quietly and peaceably returns home.

When at large no definite object possesses the animal, and he will often travel fast and far within a short time, expressive of this the term "running rabies" was formerly used. Sometimes he will return home slyly and suspiciously, but friendly with his master.

As the disease passes into the violent stage the unfortunate animal grows more restless, and on the alert, at times becomes delirious, possessed as it were by spectral illusions; springs at the door as though he heard some one approaching; again he will examine every part of his kennel or room in which confined in a most minute manner, then retire to an obscure corner to remain but for a few moments and again commence his wearisome search. At times his eyes are fixed on some imaginary insect, which with his gaze he seems to follow in its course along the walls; at last he springs forward snapping at the intruder his mind pictures, when the spell seems broken and he returns to his corner as though ashamed of his delusion.

As the disease progresses, dogs that are fastened will struggle to break their chains; those that are confined within a room will bite into the wood work, or roll about in the straw which they shake in their teeth. The violent paroxysms become longer and more severe, sometimes lasting for several hours, during which the dog snaps and bites at whatever he encounters; they seem prompted to attack other animals large and small, and less often man. Not infrequently a rabid animal will bite and lacerate his own body, even to gnawing his feet to the bone; cases are on record of terrible self mutilation; some are insensible to external impressions, and will bear blows without a cry, and bite at red hot irons.

If a stick is extended to a chained dog he will snap at it, and cling with such force that his teeth will break and his lips bleed.

Paroxysms are excited by teasing, the sight of other animals, or may occur without provocation. While they last the dog is possessed with un-usual strength, often breaking his chain and freeing himself. Bitches with whelps have been known to bite and tear their pups with their teeth.

During these paroxysms those infected seem perfectly delirious, and frequently there is observed convulsive twitchings of the face, and occa-sionally convulsions. Following the attacks the animal relapses into a sort of stupor, and if undisturbed a much longer interval will more often elapse before a recurrence of another period of excitement. In some cases the mental state is for a time so improved, masters are recognized.

A sign which is considered highly characteristic of rabies, and one which frequently makes its appearance early in the disease, is the peculiar alter-ation in the tone of the voice; the bark is described as a sound between a bark and a howl, uttered in a rough hoarse tone, which might be called croupy. The short sharp sound is often prolonged into a mournful dis-tressing wail, the animal lifting his head in the air as it issues.

The change in the voice is attributed by some to a swollen condition of the pharynx and larynx; this is possibly true and yet certain functional disturbances in the great nerve centres may be even more active, as in cases of children suffering from diseases of the brain, the cry becomes harsh, shrill, and ear-piercing.

In the maniacal stage of rabies the appetite for food nearly if not en-tirely disappears; occasionally a favorite bit if offered will be eaten but rarely is this the case. Instead of food a rabid animal will invariably swallow large quantities of indigestible substances, such as straw, clay, and dung.

Among the many ancient and unfounded traditions concerning the course of rabies, may be recalled the notions that dogs so infected froth and foam at the mouth; that they always run in a straight line, and carry their tails held closely between the hind legs, also that a peculiar odor attends a rabid dog which can be detected by one in health. All this is fanciful and unfounded; authorities state that the saliva is discharged only when the animal is unable to swallow; the tail is wagged and carried as usual until weakness causes it to drop; in running the course is much the same pursued by one uninfected The gait of a rabid dog is somewhat characteristic, described as a jog-trot. He carries his head low with tongue protruding, often torn and bleeding, swollen, and covered with dirt. Rarely will he turn from the course he is pursuing to attack anything unless it be animals of kindred species, the sight of which almost invariably excites his rage and invites an encounter.

At no time during an attack of rabies is there a special dread of water as was formerly popularly supposed. One observer is credited with hav-ing detected in extremely rare instances, dogs unable to swallow owing to spasms of the throat. Almost invariably they drink freely and deeply, showing no nervous excitement at the sight of fluids. It is not an uncommon sight to see them lick their own urine, a fact upon which much stress as a diagnostic sign is put by some authors.

As the disease progresses, emaciation rapidly supervenes, the entire appearance of the animal becomes changed to a marked degree; the head frequently swells, the eyes which sink within their cavities, are brilliant and glistening; the hair becomes rough and staring. The mucous membrane, more often dry and parched than moistened, changes to a purplish color. The breathing during a violent paroxysm and immediately after is hurried; at other times is but slightly affected.

The duration of the violent or irritative stage is generally not longer than three, or very rarely four days. As the rabid animal becomes weakened and the paroxysms grow less severe and distinct, the disease emerges into the so-called paralytic stage. The change in appearance now becomes more striking, surprisingly so considering the few days that have passed since the appearance of the first symptoms. The eyes are dim and glistening, the purplish tongue protrudes, emaciation has reduced the unfortunate to a mere skeleton; he presents a picture both appalling and heart-rending. He staggers and stumbles blindly about his kennel, until increasing exhaustion at last overpowers him; he will still bite or snap at things which may be used to arouse him, but his strength departing his efforts grows more feeble, his breathing is short and labored, his voice hoarser; he passes into a stupor, possibly into a partial or complete convulsion, and at last death mercifully closes the scene.

The progress of this terrible malady is very rapid. The different stages may be passed and death ensue on the second, or life may be prolonged to the tenth day; the latter set as the limit is but rarely reached, and from four to five days is the more common duration.

The dumb or sullen form of rabies which constitutes from fifteen to twenty per cent of the total number of cases, is but a peculiar type of the disease, which runs a much shorter course, and without the violent or irritative stage. There is decidedly less excitation of the brain, the violent paroxysms, the illusions, the constant motion, the disposition to bite, and the propensity to stray away, are all absent or present only in a slight degree, and the animal is quiet, silent, and dejected.

Paralysis of the muscles of the lower jaw is a characteristic symptom of this form of the malady, and manifests itself early in the attack. The jaw drops and the mouth remains constantly open. In rare cases a partial control of the muscles is retained, sufficient to lift the jaw and possibly allow the animal to bite if sufficiently irritated. Rarely more than a few hours, possibly three or four elapse after the disease manifests itself before this symptom appears. There is great difficulty in swallowing, and the poor dog will plunge his muzzle into the water up to the very eyes in order that he may get one drop into the back part of his mouth to cool his parched throat. It is in this form of rabies that frothing is observed, the flow of mucus and saliva in abundance dripping from the open mouth. The voice changed and of a hoarse tone, is seldom heard, and that peculiar combination of bark and howl, characteristic of the violent form of the disease, is entirely absent.

Death in the dumb form of rabies results more quickly, life being but rarely prolonged more than two or three days. The appearance of the eyes, and generally haggard and depressed look marking the derangement of the brain, the loss of appetite, the rapid emaciation and paralysis, are symptoms resembling much those seen in the violent attack.

Both forms of the disease have appeared in the same kennels, and at the same time; inoculation from dogs suffering from the one variety may give rise to either violent or dumb madness.

Diagnosis.—In some cases of rabies a diagnosis will be difficult, while in others the signs will be sufficiently pronounced to render the task easier. It is to be remembered that but few animals can compare with the dog in intelligence and high mental development, that he possesses a sensitive and easily excited nature and is extremely liable to contract nervous derangements and diseases. This fact is demonstrated by the frequent occurrence of convulsions, chorea, and kindred disorders. Again it is more than probable that the study of canine ailments will yet prove to observers the existence in dogs of mental diseases, at present overlooked and unsuspected.

When an animal presents certain symptoms which are known to appear in rabies, all, the medical men by no means excepted, are much too ready to jump at conclusions, and condemn to execution a poor animal as rabid, when he may be suffering from either meningitis, epilepsy, severe pain, excessive fear, neuralgia, starvation, toothache, parasites in the nasal cavity, acute otitis, disease of the kidneys, or some disturbance of the brain of which at present we have no knowledge. Nor must we forget the action of certain irritant poisons which when swallowed cause intense inflammation of the throat, stomach, and intestines.

Few there are who have not seen dogs in a state of delirium caused by the action of the intense heat of the sun. Others doubtless have witnessed the extreme mental disturbance of a bitch deprived of her whelps, and the distress and efforts of an animal to free himself from restraint in a new home, and return to his old master and those he loves.

Some readers will naturally observe that to mistake for rabies many of the diseases referred to must be impossible, and yet such errors are more easily made than they imagine, for not only in these but in other diseases unmentioned, there occur symptoms resembling somewhat those of the dread malady. Again it must be remembered that fear prejudices reason, dulls perception, and blunts judgement; even the thoughts of hydrophobia incite a measure of terror, and important symptoms are overlooked, while many in nowise significant become pronounced.

As one authority has said, " in forming the diagnosis, we should have constantly before us a picture of the disease as a whole, and never base an opinion upon individual symptoms, such as the propensity to bite, which may be slight, or even entirely absent."

Passing in review, the characteristic symptoms of the violent form of rabies are : the marked uneasiness, the delirium and very great excitement occurring in paroxysmal attacks, the tendency to bite, the efforts to break away, the peculiarly changed voice, the perverted appetite, the rapid emaciation, exhaustion, and invariably fatal termination.

In the sullen or dumb form the violent stage is omitted or hardly recognizable; it runs an extremely rapid course; the animals are quiet and depressed; have but little disposition to bite or run away; early in the disease are paralyzed in the lower jaw; have perverted appetite; changed voice, rarely heard; progressive emaciation and exhaustion, and seldom live beyond the third day.

Preventive Treatment. — Immediately after the bite of an animal presumably rabid, the whole wound should be sucked, and as soon as possible thoroughly cauterized. If there are no abrasions on the lips or tongue, there can be no danger whatever from the virus on the part of the one who applies suction. During the time the lips are employed, the mouth should be frequently rinsed with warm water, and the teeth used freely in gnawing as it were the edges, to keep the wounded vessels open and bleeding. To promote a flow of blood which is decidedly favorable, a cord or handkerchief can be tied fairly tight between the wounded part and the body. Suction should be persisted in until the bite can be cauterized, and discontinued as soon as that operation is possible. If the individual bitten lives at a distance from chemists, and delay must ensue before medicinal caustics are obtained, a hot iron should be depended upon, and the wound thoroughly burned.

The chemical and corrosive agents penetrate every part of the injury with greater certainty, and when possible they should be secured. The most active are nitric, sulphuric, and carbolic acids, caustic potassa, and nitrate of silver. The latter was strongly urged by Mr. Youatt who certainly proved its efficacy since he was bitten many times and escaped, though he used no other preventive; he also observed instances in which out of several animals bitten by the same dog, those which were cauterized by the nitrate of silver escaped, while some which had the wound incised, or cauterized with a hot iron were subsequently infected with rabies.

Regarding the sucking of a wound made by a suspected animal, it is of interest to recall the fact that in Lyons, during the first twenty years of the present century, certain women made it their business to apply suction to the wounds made by rabid dogs, for which they were paid ten francs for the first operation, and five for each succeeding one. Of thirty-eight persons bitten and subsequently subjected to this operation not one contracted hydrophobia.

It is yet an unsettled question whether or not it is wise to allow the wound made by a bite to heal naturally and at once, or to keep up the inflammation by applications. The latter is generally accepted as the better, and for the purpose caustic can be used every few days as the need is manifest, or a dressing of the resin ointment may be applied. In four or five weeks the wound should be allowed to heal.

It is estimated that in those cases where cauterization is resorted to about one third of the human beings bitten by rabid animals fall victims to the disease; in cases where the operation is not performed more than four-fifths of those bitten meet certain death.

In the many pages devoted to rabies the writer has pictured with wearisome detail the symptoms and phenomena which attend both forms of the disease. During the last century the public mind has been much wrought and depressed by constant allusion to the malady. Communities have often been terrorized by the appearance in their midst of dogs supposed to be rabid, and many a valuable pet has met his death from the hands of the poisoner, incited to the hellish deed through the cowardly fear of a disease so rare the dangers from it are almost nil.

Scientific research and advancement in knowledge of this subject has been so obstructed by ignorance and superstition, transmitted from the dark ages, the general public are to day as destitute of a proper understanding of hydrophobia, and are endowed with as wild fancies, absurd theories, and fanatical notions as were those of a hundred years ago.

From the British Medical Journal we quote the following on this subject.— "Something should be done to disabuse the public mind of a groundless, or greatly exaggerated terror, it would be amusing if it were not grimly sad, to observe, not unfrequently, the insane evidence of a purely mimetic morbid state set up by the misery and apprehension caused by the consciousness of having been bitten by a mad dog. As a matter of sober medical fact, it is by no means necessary or inevitable that the bite of a dog with rabies should give a man or woman hydrophobia; and if the element of fear could be eliminated, it is highly probable that the proportion of instances in which the dreaded disease supervened from a bite would be greatly reduced."

Many pages have already been devoted to this subject and much of equal and even of still greater interest is to follow. The task has proved herculean, but the writer has felt his duty imperatively demanded that an exhaustive description of the symptoms and phenomena of rabies be pictured by him to the reader. Although he has never been able to appreciate why people are possessed by that insane terror of a disease, so rare that death by the hand of the executioner might with equal good sense be anticipated, he has felt that superstition, and an ignorance of the subject were conditions favorable to the propagation of the common terror, and only by enlightenment born of a perfect understanding of the disease and its actual not imaginary dangers, can the public mind be disabused, and the groundless terror dispelled.

In pursuance of the author's intention to familiarize the reader with all the more important facts developed by scientific research, he submits the following paraphrase of the address of M. Pasteur at the International meeting at Copenhagen, August, 1884. —

"Congresses seem to point out the chief directions of progress, whilst they are also meetings for examination of the most important problems in medicine.

When three years ago, this Congress met in London the microbe theory was still in attack but now it is admitted, and our opponents should acknowledge it. That spontaneous generation was a chemical hypothesis has also been proved; on the other hand microscopic life has a relation to organic decomposition and fermentation, hence theories as to spontaneous origin of disease must cease to exist.

From the London Congress another advance must also be dated viz., the possibility of attenuating viruses, varying their infectious nature and preserving them. The application to veterinary medicine followed, and animals are protected against such diseases as fowl cholera and splenic fever. The methods applied to animals might be adopted in regard to man only we should have to proceed with caution. The inquiries hitherto conducted are things of yesterday; they have been fruitful in results and we have a right to expect further advances. I was induced to study rabies, a disease surrounded with obscurity, in order to penetrate further into the arcana of knowledge. For four years in my laboratory, we have been conducting experiments under difficulties, though the French government has in its zeal for scientific interests endeavored to smooth the way. I shall to day give the results of our experiences.

Every disease and especially rabies makes us think of a cure but in searching for a remedy we often indulge in fruitless labor. By studying the nature of the disease and development, we shall more certainly lead to the remedy.

It is proved that the virus of rabies always develops in the nervous system, brain, spinal cord, nerves, etc., and never invades at once all these parts. It may conceal itself in the spinal cord and then attack the brain or vice versa. We find that the uppermost part of the spinal cord, which is called the bulb, is invariably the seat of the poison, so that when an animal dies of rabies we are able to obtain from this bulb, some of the virus of rabies which will produce the disease, if inoculated on the surface of the brain. You can test this on any dog of any kind by trephining. The experiment never fails and has been performed by us numerous times on hundreds of animals.

The invariable existence of the virus in the bulb of dead animals from this disease, the certainty of communicating rabies to other animals by opening the brain are truths now established. As to the origin of hydrophobia or rabies, all forms came originally from the bite of a mad dog. Rabies never rises spontaneously in the dog or other animals; all cases to the contrary are non-authentic.

There must have been a first case of rabies, but this touches problems unsolvable. It touches the problem of life. Who would maintain that an ovum always came from an ovum and that the first ovum spontaneously developed? Science does not argue about the origin of things. No one is benefited and such questions are beyond its province.

Nothing is more variable than the natural period of incubation in this disease. One dog goes mad in four or six weeks, another in two or three months. By one method of inter-cranial inoculation, the period of incubation is shortened and known approximately. The duration of incubation may also depend upon the quantity of active virus reaching the nervous system unchanged. The quantity of virus may be infinitesimal.

On May 10th., 1882, ten drops of a fluid, obtained by macerating, in a sterilized broth, a portion of the bulb of a wandering mad-dog, were introduced into the popliteal vein of a dog. A second dog was inoculated with one one-hundredth, and another dog one two-hundredth of the quantity. The first dog had rabies on the eighteenth day, the second on the thirty-fifth, and the third was unaffected, the quantity of virus not being sufficient. This dog was again tested and contracted rabies twenty-two days later. If we operate by trephining we obtain a method certain in its effect. If we take mad dogs at any season, and in each case isolate the bulb and inoculate with the material by trephining a few rabbits, the phenomena will be regular. No matter what dog be used the incubation will fall within twelve or fifteen days; it will never be eleven, ten, or eight days though it may be over fifteen days. Other instances of the peculiarities of the virus might be given, all proving that there exists and can be produced different kinds of rabies all more or less violent and fatal.

Guinea pigs soon attain a maximum of virulence; the incubative period is shortened, and by transmission we obtain a virus which surpasses in virulence that of rabies ordinarily met with.

Jenner propounded the theory that poison called "grease" in the horse, now described more correctly as the horse-pox may be weakened in activity by transmission through cows. This idea of Jenner's struck us as being capable of imitation. Could we attenuate the virus by passing it through the bodies of certain animals? Many attempts were made, until in the case of monkeys, a medium was found.

December 6, 1883, the bulb of a dog which had been bitten by a child who died of hydrophobia, was taken and a monkey inoculated by trephining. In eleven days the monkey became rabid. From this monkey, the virus was transmitted to a second. In eleven days the second showed signs of rabies. A third monkey became rabid in twenty-three days. Other monkeys were experimented on. With the bulb of each of these monkeys two rabbits were inoculated by the method of trephining. The rabbits infected by the first monkey became rabid on the thirteenth and sixteenth days; the two from the second monkey on the fourteenth and twentieth days; the two from the third monkey on the twenty-sixth and thirtieth days; two from a fourth monkey on the twenty-eighth day in each case; two from a fifth monkey on the twenty-seventh day in each case; two from a sixth monkey on the thirtieth day in each case.

By transmission from monkey to monkey, and from monkey to rabbits, the strength of the poison seemed weakened. A dog inoculated with the bulb of the fifth monkey had incubation period of not less than fifty-eight days; other experiments confirmed these results. We thus found out a method of attenuating the virus, and a method of vaccinating dogs as prevention against rabies.

As a starting point we take one of the rabbits inoculated from monkeys to such a degree, that hypodermic or intravenous injection does not cause death. The preventive inoculations are done with the bulbs of rabbits which have been successfully infected from monkeys to rabbits. Jenner's methods met with opposition. Remembering this I determined to lay my results before a scientific commission. M. Tallieres, Minister of Instruction in France, supported my project. M. Beclard, P. Bert, Bouley, Tisseraud, Villemin, and Vulpian were appointed to examine my statements and check the facts I had communicated to the Academy of Sciences May 29, 1884. M. Bouley was chosen President, and Villemin, Secretary. The committee have recently presented their report to the Minister. I am now able to give a brief account of the first report.

I presented to the committee nineteen vaccinated dogs, all of whom had been rendered insusceptible by preventive inoculation, and thirteen of these had been further tested after vaccination by inoculation by trephining.

These nineteen dogs were compared in various ways with nineteen other dogs chosen for the purpose of experiments. On June 1st. two of the protected dogs and two trial dogs were inoculated by trephining under the dura mater with the bulb of a mad dog. On June 3d. one protected dog and one trial dog were bitten by a mad dog. On June 4th. the same mad dog bit another protected and another trial dog. June 6th. the mad dog used June 3d. and 4th. died, and the bulb was inoculated by trephining on three trial and three protected dogs.

June 10th. one protected and one trial dog were bitten by another mad dog obtained from the streets; June 16th. one protected and one trial dog were bitten by a dog which had gone mad June 14th. as a result of the experiments; June 19th. three protected and three trial dogs inoculated in popliteal veins with bulb of mad dog; June 20th. six protected and four trial dogs inoculated in vein; June 28th. two protected and two trial dogs bitten by mad dog at the hospital of M. Paul Simon, a veterinary surgeon.

Now let us look at the results.—The Com. performed experiments on thirty-eight dogs, nineteen protected and nineteen not protected. They report that in the case of the nineteen trial dogs, rabies occurred as follows:—of six bitten, rabies occurred in three; of seven inoculated in vein, rabies in five; of five trephined, all died. Trephining was the surest. On the other hand, not a single symptom of rabies appeared in any of the dogs vaccinated by me and declared insusceptible; one dog died from diarrhœa, the cause of death being verified by post-mortem and also by tests. Three rabbits and one guinea pig were inoculated with the bulb of this dog and are in the best of health. The animals are still under observation."

The following is a literally translated extract from the address of M. Bouley, before the French Association for Advancement of Sciences.—

Living nature of the contagion and Inoculation preventive of the Hydrophobia.

"Nothing is conjectural now. The agent of the contagion is a living agent, already known for an important group of contagious diseases. This agent, this living element, we have been able to master, to observe it in the midst of artificials proper for its culture; we have been able to study its physiology, modes of existence; we have been able to subject it to profound modifying influences; to create, so to speak, races in different species. In short we have been able to see it by our own work in the living organisms, in conditions vigorously determined by the experimenters, and to study the series of effects that it is capable of producing.

After having torn away from nature this secret, so long a time guarded, that contagion is the function of the living element, that a certain number of contagious diseases is possessed of this element, they have isolated it from organisms where they had taken it, they have studied it in the midst of artificials proper for its development and to all the manifestations of its life; have submitted it to influences capable of moderating its energy, and in short have resolved this wonderful problem of transforming it, it the agent of death, into an agent of efficacious preservation which renders invulnerable those attacked by the natural contagion of the animals which have been submitted to the influence, mitigated by a systematic culture of the living element of this contagion; and this grand discovery of the attenuation of the virus that the entire world has applauded as one of the most wonderful acquisitions, the most marvellous that medical science has ever made perhaps; this discovery which in the capital of Scotland some months ago, and some days ago in the capital of Denmark, has yielded to M. Pasteur one of the grand triumphs which were formerly accorded to the conquerors in war.

Hydrophobia from the beginning of time to these last hours has lived a fatal disease, against which all attempts have ever and ignominiously failed, and here to day, thanks to M. Pasteur and his collaborators that he has associated in his work, it can be transformed into a disease wholly benignant, which not only is compatible with life, but still has this happy privilege of rendering invulnerable the organisms which have received the germ, in that state of benignity, against those harmed by its terrible virus, when they are inoculated in the natural conditions of intensity. To render hydrophobia as harmless a disease as carbuncle has become, what a marvellous problem is solved!

They knew positively that it was contagious, and that its exclusive mode of transmission was inoculation, which might be accidental, like that which results from a bite, the most frequent of all; it may be inoculation by the deposit of the foam of a mad animal upon an absorbing mucous, or wounded surface, or may be in short, experimental inoculation.

When they inoculate the monkey in successive series, the energy of the virulence follows a descending scale, of which the degrees can be measured by the increasing duration of the period in inoculation, and the attenuated hydrophobia of the monkey is transmitted to the dog or rabbit; the duration of the period of inoculation with these last animals is witnessed by the proportionate long increase of the attenuation of the virus which was inoculated to them. The application of these facts says M. Pasteur places in our hands a method of vaccinating dogs against hydrophobia.

Messieurs.—This remedy, after which the victims of the bite of rabid animals have always looked forward to with so much ardor, the day is not far off without doubt, when M. Pasteur and others will be able to place it in the hands of physicians, and this remedy which is a very paradoxical thing, this will be the terrible virus itself, removed by the wonderful method of attenuation of its deadly energy, will have been transformed in a vaccination against itself, and will be able to invest the organism to which they inoculate it with a benificent immunity, thanks to which the natural virus, inoculated by the bite, can no longer take effect on it."

M. Bouley sums up as follows.—

HYDROPHOBIA AND BENIGNITY!

HYDROPHOBIA AND IMMUNITY!

HYDROPHOBIA AND CURE!

CONVULSIONS.

Convulsions and fits are terms used synonymously and signify a sudden seizure, loss of consciousness, involuntary, spasmodic contractions of the muscles of the body, foaming at the mouth, followed by stupor.

Convulsions among dogs, and more especially in early life, are of frequent occurrence. Epilepsy is the more common term used among writers on canine diseases to designate these attacks; it should more properly be restricted to those purely functional in character, that is, not necessarily attended with either inflammation or appreciable disease of structure, and distinguished by a chronic course and unexpected recurrence.

Convulsions do not represent any single form of disease. Among the number of morbid conditions which may induce them some are determinable, others are obscure, and of them we possess little positive knowledge.

Causation. — Predisposition plays a prominent part in the production of numerous forms of convulsion; it may be inherited, or acquired through the action of various conditions and diseases.

Nervous disorders are more easily generated In some breeds of dogs than in others, and in-breeding among such cannot lessen the liability. General debility, changes in the blood, age, and diseases incident to parturition, appear strongly to favor a predisposition.

Among the exciting causes may be mentioned worms, the eruption of the teeth, over exertion, intense heat, fear, anger, and witnessing other convulsions. Prof. Dalton has related the following illustration of the effect of imitation in the canine race. A dog, not previously affected with epilepsy, was in company with another dog who was subject to the affection; the latter being seized with a convulsion, the former immediately afterward had a similar attack.

Among the causes exciting convulsions may be mentioned numerous poisons, strychnine etc.; injuries to the brain and spine; in fact very many irritations and inflammations in different parts of the body, and derangements and diseases of the internal organs may give rise to the affection.

Symptoms. — Convulsions come on abruptly and without symptoms indicative of their onset. If the dog is at exercise, he stops suddenly, remains for an instant fixed to the spot as it were, his legs tremble, uttering a short, sharp cry or a low moan he falls, possibly attempts to rise again, again falls, becomes unconscious, and convulsive movements at once begin. More often they are at first what are termed tonic spasms, that is a persistent and uniform muscular contraction of great intensity, lasting usually for a few seconds, followed by spasms for which the name clonic is given. By this it is understood a rapid succession of contractions and relaxations of the muscles, by means of which the body is kept in a state of continual and very active movement. The head, the limbs, and the body jerk violently; the tongue is sometimes caught between the teeth and bitten; the movements of the jaws are accompanied with a foamy saliva, frequently tinged with blood from the wounded tongue. Respiration during the tonic spasm is arrested, and is irregular and incomplete while the active convulsion exists; in consequence the mucous membrane becomes livid and congested.

A convulsion may be slight or severe and prolonged. As the paroxysm ends the animal draws a deep, long sigh, and if the attack is epileptic he then soon recovers consciousness, and is moving about as though nothing unusual had occurred. In some cases the unfortunate passes into a state of profound stupor, in which he remains a variable period. Gradually consciousness returns, he makes several attempts to rise, reels about for a few steps in a bewildered manner, grows steadier, and is soon himself or much improved. In very rare cases he will appear delirious, and rush wildly away or towards those about him.

Prognosis. — Since convulsions cannot be considered as representative of any single form of disease, and they relate to many morbid conditions differing from each other so widely, unless the cause is known which induces them, a prognosis cannot be made.

In epilepsy the paroxysms recur after intervals extremely variable. When worms cause an attack, their removal promises a cure. The so-called suckling fits are due to exhaustion and disappear when strength returns. Convulsions occurring in distemper from brain irritation almost invariably prove a fatal symptom. The same may be said when they appear in disease of the kidneys, and as the result of poisons. Teething fits sometimes end fatally but are more often recovered from.

In a general way it may be said the prognosis is favorable when the causes are of a simple nature and easily removed.

Treatment. — During a convulsion the animal should be secured to prevent his running away if delirious on recovery. The object of treatment then is to cool the head and keep the body warm, thereby drawing the blood from the brain. Bathing the head with cold water will generally prove quite sufficient. After the convulsive movements cease if he is inclined to sleep, allow him to remain undisturbed.

In every case the primary cause should be sought for, and efforts be made to effect its removal. If poison is suspected, proper treatment will be found advised elsewhere. If the convulsion is prolonged, the injection of the hydrate of chloral recommended in strychnine poisoning should be given. Generally where spasms are connected with diseases or derangements of different organs, the existence of such disorders will have been manifested by previous symptoms, and their cure is demanded.

The treatment of general predisposition should be tonic and applied to the nervous system. The oxide of zinc in two grain doses twice daily is a reliable agent, and should be persisted in to secure a full effect. Over-feeding, plethora, want of exercise, worms, etc., suggest a change in management and the needed medication.

VERTIGO.

Vertigo or dizziness is an occasional symptom among dogs. If severe they fall suddenly, remain motionless for a few moments, and are again moving about as before. In mild cases they simply reel for a few steps like one intoxicated. At times vomiting occurs. Generally the cause can be attributed to indigestion or liver derangements. Sometimes over-eating, and occasionally deprivation, are active influences. A tight collar may interrupt the circulation in the head and therefore induce an attack.

Treatment. — The most important measure is to obviate the cause. After that is removed tonic remedies, nutritious diet and other means of improving the general health are indicated.

APOPLEXY.

The term apoplexy signifies a sudden stroke or shock. It does not represent an individual disease, but is applied to sudden lethargy or profound stupor, occurring in different affections of the brain.

Causation. — The most frequent condition which gives rise to apoplexy is hemorrhage within the skull, the blood pressing upon the brain substance thereby inducing paralysis. Sometimes an apoplectic attack is induced by sudden congestion of the brain, as in cases of sunstroke. Another condition which may give rise to it, is an interruption in the circulation of blood in a portion of the brain by the plugging of an important artery. This accident is caused by clots, which may have formed either in the heart or arteries. Profound stupor may occur in inflammation of the substance of the brain, and as the result of uremic poisoning in diseases of the kidneys.

Certain morbid conditions within the skull may exist, and other influences which invite an unusual accumulation of blood there, may co-operate to cause an apoplectic seizure. Great mental excitement, anger, intense heat, violent exercise, blows on the head, and straining are among the exciting causes. A short neck and full habit, as seen in certain breeds, was formerly concluded as denoting a predisposition to apoplexy; this supposition is not sustained by facts.

Symptoms. — The abruptness of the attack, partial or complete unconsciousness; slow, heavy, noisy breathing; full, hard pulse, less frequent than in health; eyes fixed and reddened; mucous membrane purplish; convulsive movements observed in some few cases. These are the more common symptoms in apoplectic seizures.

Treatment. — The head should be raised and kept cold by applications, and the body warm by coverings and artificial heat if needed. Generally a cathartic should be given as soon as the animal can swallow. Too active measures are unwarranted; bleeding can do no good, and the treatment must be expectant; perfect rest, and efforts to sustain the vital powers by proper nourishment are the great essentials. Nature must be aided and depended upon. Apoplexy is a grave condition; recovery may take place, and the chances of a recurring attack depend upon the cause which gave rise to it in the first instance.

CONCUSSION OF THE BRAIN.

Concussion signifies sudden shock and interruption of the functions of the brain. The condition is rare among dogs, but may exist, caused by a blow or other mechanical injury to the head.

Symptoms. — In ordinary cases the animal lies for a time motionless and unconscious. If an attempt is made to arouse him he opens his eyes, moves slightly, and again becomes insensible. After a time in this state, the functions of the brain renew their activity, he moves restlessly, generally vomits, and his senses return.

In more severe cases of concussion, the animal is unconscious of efforts to arouse him, his breathing is slow, and his pulse quick and feeble; surface and extremities cold. Vomiting in these cases rarely occurs. As reaction goes on the pulse grows stronger, the breathing quicker, circulation better, and the body warmer. After a time partial sensibility returns, and when at last on his feet he staggers, reels blindly about, and possibly again becomes unconscious.

Death may follow a severe concussion; in some instances recovery takes place slowly; in others it is only partial, the animal remaining infirm, with intelligence lessened.

Treatment. — The skull should be carefully examined for a possible fracture, and if one is found a surgeon should be immediately consulted. In mild cases there is but little need of active interference. When the symptoms indicate the injury a severe one, cold should be applied to the head, and heat to the body, to lessen the flow of blood to the brain, and thereby prevent inflammation.

As soon as the animal can swallow, a purgative should be administered, and his bowels subsequently kept active. If inflammatory symptoms manifest themselves, leeches should be applied to the back of the head and followed by a blister. Perfect quietude is a marked essential. The diet should be mild and unstimulating.

The effects of a severe concussion often persist for a long time and may eventually indicate the need of strychnine in small doses.

HYDROCEPHALUS.

This affection is occasionally seen among dogs in early life. The formation of tubercules on the membranes of the brain, is generally stated to be the exciting cause; the normal fluid contents become intensely augmented, and portions of the brain degenerate and soften. The affection is more often present at birth, but is not necessarily congenital. A state of impaired nutrition, a weakly constitution, and the co-existence of rickets or scrofula is the more often noted.

Puppies affected with the disorder show a lack, and possibly a total absence of intelligence. Very often partial paralysis of the hind legs occurs, and the dull, drowsy, symptoms observed in compression manifest themselves in a greater or less degree. Convulsions often set in late in the disease. The general appearance of the animal after a time clearly indicates the hopelessness of treatment.

COMPRESSION OF THE BRAIN.

Compression of the brain is more often produced by either an effusion of blood, tumors, fracture of the skull with depression, or by the formation of pus or serous fluid within the cranial cavity.

Symptoms. — Partial or complete unconsciousness and stupor, breathing slow, labored, and noisy; pupil dilated and insensible; pulse slow; sometimes the urine dribbles away, and a discharge of feces occurs involuntary.

Diagnosis. — To discriminate between concussion and compression of the brain is by no means easy in some cases. The leading points of distinction are as follows:— The symptoms of concussion appear immediately after the accident; those of compression from an effusion of blood, may manifest themselves after an interval. Be it understood that for a time after a severe blow on the head it will be difficult, if not impossible, to tell whether simple concussion or compression exists. If the first symptoms disappear readily, and the animal seemingly having entirely recovered afterwards becomes unconscious, it may be inferred that compression has followed concussion. In the latter the pulse and breathing is weak and feeble, the body cold; in the former the breathing is heavy, labored, and noisy; the pulse slow and full; complete muscular paralysis is not uncommon in this condition, but rare in concussion.

Treatment. —Fractures demand surgical treatment, the same may be said if it can be determined that pus has formed within the cranial cavity. In other respects the treatment advised in apoplexy should be employed in compression.

ACUTE MENINGITIS.

Acute inflammation of the membranes of the brain is an affection occasionally seen among dogs, and deserves especial consideration from the fact that there is presumptive evidence that many cases are taken for rabies, from which in a certain stage it is not always easy to discriminate.

Causation. — The usual causes are traumatic, cerebral concussion, injuries acting directly upon the substance of the brain, intense cold, sunstroke, great mental excitement, and the extension of inflammation from adjacent organs. The disease may take its origin in otitis; less frequently it extends from the cavities of the eyes. Meningitis is sometimes developed in connection with distemper, and may follow other constitutional diseases.

Symptoms. —The affection in rare instances comes on gradually but more often the onset is abrupt. The first stage is one of active congestion, and is characterized by pain, delirium and maniacal excitement. It is now that the disease is often taken for rabies. The behavior becomes changed, the voice altered, the eyes are glistening and vacant in expression, the animal is extremely restless, snaps and barks at imaginary intruders, and bites at sticks extended to him.

In meningitis as in that dreaded madness the violence and maniacal excitement occurs in paroxysms, with intervals of comparative quiet. Convulsive movements occur, snapping the jaws, champing the teeth, and foaming at the mouth, and very often prolonged general convulsions are observed. The bowels are constipated. The sense of hearing remains acute, and when called the animal will raise his head as though listening, but unable to locate from whence it comes or appreciate the significance of the sound, he does not respond.

When on his feet, if confined in a room, he circles around it sniffing at the walls, at times stopping and barking for several minutes, his voice altered and the tone pitched high. His legs tremble under him showing increasing weakness. Liquids he drinks readily and with feverish rapacity. Vomiting frequently occurs; the eyes are bloodshot, the face haggard; the pulse is quickened, and the temperature of the body notably raised. Restlessness is a marked symptom in this stage. The writer recalls the case of a young mastiff afflicted with the disease, which he allowed to run at large within his house, that the symptoms might be carefully observed by the inmates. This dog while he had strength to make the distance had a certain point to which he invariably journeyed. He would start from his bed in the kennel man's room, climb three stairs, enter the kitchen, pass from there into the dining room of his master, and stop at the hall door, then without pausing, take the same direct course back, and on reaching his bed again turn and travel the same distance, always for the same points, never deviating or passing through other doors or going beyond his self imposed limits. Other dogs lay about the rooms undisturbed and unnoticed. At first his journeys were made on an easy walk, head carried low; later on he entered a run which he kept up for an hour at a time until exhaustion overcame him, then for a brief interval he remained quiet, and when his strength returned he would again start on his wearisome run. The inmates of the house would occasionally meet him on his journeys; without any disposition to bite he would deviate only sufficiently to pass them and continue on. For two days only had he strength to climb the stairs, but until he died, some three days later, he constantly made feeble efforts to do so.

The stage of active congestion in dogs suffering from acute meningitis is short, rarely more than two or three days, when the symptoms change as an effusion forms within the cranial cavities and presses upon the brain. Drowsiness succeeds the maniacal excitement; the sight becomes dim or wholly disappears; obstructions are no longer avoided but blindly encountered. In his movements the animal seems wholly unconscious, crazed as it were; his bark is lower and feeble; he still drinks if his nose is guided to the basin; the intervals of quiet grow much longer; he rises to his feet with difficulty, his legs weak and trembling; the stupor grows more profound; paralysis ends his tiresome walks; convulsions occur often and severe; death finally results.

The disease under consideration may run a fatal course in two or three days; rarely will it extend beyond six or seven.

Diagnosis. —From the foregoing symptoms it will be understood how easy it is for an observer, unfamiliar with the manifestations in both diseases to mistake acute meningitis for that dread malady, rabies, and yet after a careful analysis of both diseases the difference is as easily appreciated. The great danger of confounding them lies in the anxiety which almost invariably posesses people to sacrifice dogs, on the first appearance of symptoms which indicates the barest possibility of rabies; they jump to conclusions and a diagnosis without sufficiently studying the case.

To kill an animal suspected of being mad, is not the first but the last thing to do. He should be secured and every possible precaution instituted to prevent injury to those around him, and then be carefully watched. The wisdom of this plan is all the more evident if the animal has bitten anyone; many a mind has been nearly crazed, by days of terror and horrible expectancy which might have been averted, had not panic stricken friends hurried the poor dog out of the world, when he should have been allowed to die naturally from meningitis, or possibly some other disease, in the course of which a few symptoms of rabies appeared.

The manner of the attack will aid much in diagnosis. Acute meningitis occurs more often after an accident, injury, some unusual exposure, or is developed in connection with some other disease. There is no melancholic stage as is seen in rabies, no shrinking from strangers; the disposition to worry articles, carpets, chair legs, etc., to eat indigestible substances, to lap urine, cold stones, and iron, to stray away, to attack other dogs, is absent in meningitis. Again while the voice is altered, the bark is short, sharp, high in pitch, entirely unlike the hoarse, croupy, blended howl and wail heard in rabies.

In that dread malady wood work is bitten, straw shaken in the teeth of animals infected; sticks extended are held savagely in the mouth and withdrawn only with great effort. In meningitis these symptoms are absent; an animal ill with the disease will bite at a stick extended, but almost immediately relinquish it, and another important diagnostic difference is, while maniacal excitement occurs in paxoxysms, it cannot be induced by worrying the animal as in the case of madness.

In the latter disorder there appear peculiar illusions; the unfortunates will see as it were, bugs, spiders, and the like crawl along the walls, and follow them with their eyes in their imaginary course; this symptom does not appear in meningitis; neither is there a disposition to bite other animals as is the case in rabies. In the former affection vomiting generally occurs, and what is of great importance to consider, is the fact that it is an inflammatory disease, and the febrile movement is more or less marked.

Prognosis. — Acute meningitis is a grave disease and recovery will but rarely take place. When developed in connection with other disorders as distemper, a fatal result may be anticipated.

Treatment. — An animal manifesting the symptoms should be secured, and a measure of anxiety will be removed if care in handling is observed.

The timorous can use heavy buckskin gloves, and will doubtless feel safer for it, although after meningitis is positively determined such precautions are needless. Perfect quiet is to be enforced; the diet should be liquid bland, and easily digestible, given often and in smaller quantities if vomiting is persistent. The bowels should be kept active; the syrup of buckthorn in tablespoonful doses, is an easy cathartic.

The convulsions demand an anti-spasmodic remedy, and belladonna is advised; the tincture can be administered in five drop doses every three or four hours as needed.

When the stage of effusion is reached, the iodide of potassium should be given in three grain doses, four times daily, and a blister be applied to the back of the neck. As the disease progresses the need to support the powers of the system becomes more urgent.

CHRONIC MENINGITIS.

Inflammation of the membranes of the brain may take on a sub-acute or chronic form. It is a disease which comes on insidiously, without characteristic symptoms, and its presence is usually for a long time overlooked, the manifestations being attributed to some other morbid state. To ascertain the cause in many cases is no less difficult than to reach a correct diagnosis; injuries to the head are doubtless largely instrumental in inducing the affection. Chronic meningitis is one of the diseases upon which epilepsy is dependent; it is possible that in some diseases of the kidneys it occasionally takes its origin.

Certain diseases within the cranial cavity may exist for a considerable period without manifestations clearly indicating the head as the seat of the derangements; even when located the absence of distinctive symptoms renders it impossible to distinguish chronic meningitis from many other affections of the brain, and a positive diagnosis can rarely be made excepting at the autopsy.

If a dog's disposition changes; if he gradually grows dull, stupid, and disinclined to exertion; sleeps more than usual; has occasional convulsive movements, possibly confined to the mouth; carries his head low; his eyes become vacant in expression, dim in sight or sightless; his movements when on his feet are erratic, and it is known that sometime in the past he has experienced a severe blow on the head, then the presence of chronic meningitis may with reason be suspected. The symptoms described are common to a variety of affections of the brain, but when following an injury the membranes are more often affected than the brain substance.

The treatment consists of mild laxatives, blisters or setons to the back of the head, perfect quiet, nutritious diet, the iodide of potassium in three grain doses three times daily, and possibly after this agent has been used for several weeks, strychnine can wisely be substituted. The appetite and general condition of the animal will indicate whether quinine and iron need be added to the strychnine.

CHOREA.

Chorea or as it is more commonly called, St. Vitus' dance, is a spasmodic affection generally accepted as purely nervous.

The causation is obscure. Worms have been assigned, but the evidence is insufficient. Great mental excitement, fear, and anger might induce the disease. It very commonly follows distemper. It is characterized by irregular contractions or twitching of certain muscles, movements which the animal has no power to control. The affection may be extensive, involving several parts of the body, but it is more generally confined to one set of muscles, the head or a limb.

Treatment. — The disease is exceedingly obstinate, and recovery is by no means certain. The purpose should be to remove if possible all morbid states of the body which may tend to aggravate the disease, such as constipation, indigestion, worms, debility, etc. To sustain the general strength and improve the vigor of the nervous system by tonics, nutritious diet, and every influence conducive to perfect health.

A diversity of remedies have been advised by different authors as a cure for this disease. Doubtless the most serviceable drugs are strychnine and arsenic; the preference given to the former. It should be commenced early in the affection and may be given as follows.—

℞ Strychniæ Sulph.   gr. ij
     Aquæ   ℥ i
Ft. Mist. Sig. Dose three drops twice a day.

The dose should be increased one drop daily until the physiological effects of the strychnia are observed, as evinced by a stiffening of the legs or neck. When this occurs the dose should be at once reduced to three drops, and increased again as before. It will not do to make the giving of this medicine a matter of convenience; it must be faithfully and regularly administered, and its action carefully watched. If no decided improvement follows the use of strychnine after a reasonable quantity has been given, arsenic should be substituted. The most convenient preparation is Fowler's solution and of this four drops can be given twice daily, and increased one drop every two or three days until the specific effects of the medicine are produced when it must at once be discontinued. These are puffiness about the eyelids, loss of appetite, a disturbed digestion, occasionally a falling out of the hair, and salivation. If it is deemed wise to add iron to the arsenic, one half an ounce of the ferri et ammoniæ citratis can be mixed with an ounce of Fowler's solution. The dose of this preparation is three drops. If the bowels are constipated, the syrup of buckthorn should be given as needed to promote a free action. In exercise, fatigue should be avoided. The use of the nerve tonics, strychnine or arsenic should be prolonged even after the chorea has disappeared, and when they are discontinued, cod liver oil in tablespoonful doses should be substituted, and persisted in until perfect health is restored.

NEURALGIA.

Neuralgia is a disease of the nervous system manifesting itself by pains. Although its presence among dogs is rarely detected owing to the difficulties clearly apparent, it is presumed that the affection is not an uncommon one, and the symptoms of it are misinterpreted and attributed to some other disorder. An ancient author thus defines the disease, "neuralgia is the cry of a nerve for more blood or better blood."

Many exciting causes both local and constitutional are recognized as active in producing the affection. The former by pressure or other influence applied directly to the nerve itself; the latter by morbid states of the blood, exposure, decayed teeth, disorders of digestion, debility and many other derangements.

Neuralgia may appear in any organ or part of the body supplied with nerves of sensation. The pain usually manifests itself abruptly, and in severe cases is excruciating, forcing an animal so attacked to howl piteously, and act in an insane manner while it lasts. It usually occurs in paroxysms with intervals of comparative ease.

By careful examination and patient watching the region in which the pain exists may be determined, and the cause should be removed if possible. Decayed teeth demand extraction, general derangements the use of tonics, local disturbances the proper remedies to overcome them. If the affected nerve can be located and is accessible, a blister should be applied, and dressed with a little extract of belladonna.

If the pain is excessive and attacks are prolonged, laudanum in twenty drop doses is indicated. If the pains occur regularly each day or night at about the same hours, three grains of quinine should be administered four times daily until they cease.

PARALYSIS.

Paralysis is a symptom, not a disease. It is characterized by impairment or loss of muscular contraction whilst the power of making an effort of the will remains. It may be partial, confined to one or more muscles; or the muscular system may be involved.

Paralysis may arise from diseases of the great nerve centres, the brain and the spinal cord; it may be due to certain influences which diminish or destroy the integrity of the nerves themselves; it may result from a diminution or loss of excitability and contractility of the muscles.

Causation. —Contusions, wounds, fractures, and other injuries produced traumatically. Diseases of the parts in the region of the nerves, thereby causing compression, as in the case of tumors, enlarged glands, and many other conditions.

Diseases of the brain and spinal cord are amongst the most frequent causes of paralysis. Disturbances in circulation as in a clot plugging an important artery in the brain, thereby cutting off the supply of blood from a portion of it. Certain poisons are known to produce paralysis. The affection in some instances occurs after acute diseases; it appears in exhaustion of the nervous system.

In dogs is often observed the so-called reflex paralysis, occurring in consequence of some disease, injury, or irritation, existing at points of nerve distribution, as in intestinal disorders and affections of the kidneys, bladder, etc. Worms, constipation, and contusions not uncommonly cause an impairment or loss of power in the hind parts.

Symptoms. —If paralysis is partial it is shown by a trembling, feebleness, and uncertainty of the movements of the parts affected. If complete, the muscles are more often relaxed and incapable of the slightest resistance. In spinal paralysis the uniform affection of both sides is the more common feature. When the brain is the seat of the disease, loss of power almost always occurs on one, and the opposite side of the head or body. When the affection is reflex, the impairment or loss is limited to the region supplied by one or a few nerve trunks.

The disease may come on suddenly or gradually. After existing for a time the muscles lose their contractility, waste, and power of motion in them is permanently destroyed, or the affected nerves may change, become incapable of conducting impressions, and thus recovery is impossible.

Treatment. —In cases of paralysis occurring suddenly, the bowels should be moved freely, and the ability of the animal to void his urine be determined by watching. If the bladder is involved by the disease it must be emptied by a catheter. It is of the greatest importance that the cause be determined and treatment directed to its removal.

In reflex paralysis induced by intestinal troubles, worms, etc., proper remedies are advised elsewhere.

Poisons demand their antidotes, and loss of power occurring after acute diseases suggests improved nutrition and nerve stimulants. If the impairment of power has been gradual, and the disease is evidently chronic in character, recovery is doubtful.

An important indication in treatment is to improve the general health of the animal as well as to restore power in the paralyzed parts. The kennel arrangements should be unexceptionable, the diet generous, digestible, and sustaining. Natural movements of the muscles should be executed daily, and if walking is possible, gentle exercise is imperative.

The medicinal agent to be depended upon as a nerve stimulant is strychnine, and should be given in gradually increasing doses as advised in chorea. Electricity is of great benefit in some cases, and should be used to assist other means employed.

While strychnine is the important remedial agent, others should be administered if their need is manifest. Quinine, iron, and cod liver oil are demanded in many cases.

Tetanus.

Tetanus is a disease manifested by continuous muscular spasm or rigidity. In the majority of cases this affection is caused by a wound or local injury of some kind. Exposure to cold and intestinal disturbances may induce it; in poisoning by nux vomica and strychnia the convulsions are tetanoid. The attack may be general or partial; when partial it is mostly confined to the neck and jaws, and hence the affection is known as lockjaw. The jaws are firmly shut by the rigid contraction of the muscles, and the mouth cannot be opened by any force it would be prudent to employ. The angles are drawn and the lips are bathed in frothy saliva. Often swallowing is impossible, owing to the throat being involved.

The muscles of the eyes and face are at times affected, giving rise to hideous distortions. When tetanus is general the body and limbs are stiff and immovable, the abdominal walls shrunken and rigid; respiratory movements of the chest are restrained. The muscular spasm is persistent; but it occasionally subsides a little and then comes on again in fits of greater violence.

Death may result from exhaustion, the nervous system being worn out by the violence of the spasm, or from suffocation, respiration being too long suspended during a violent convulsion. In very rare instances recovery takes place on the removal of the cause.

Treatment. — To remove the cause is the first axiom in the treatment of all diseases. Other measures to be employed are to relax the spasms and sustain the strength of the animal. Perfect quiet is to be enforced and the administration of remedies should be effected as gently as possible. When difficulty in swallowing exists, nourishment and medicine should be given per rectum.

A wound if apparently the cause of the disease, should be reopened and a free discharge of pus promoted; cauterization will be necessary if it is ragged and ill-conditioned, and afterward poultices ought to be applied.

Among the many sedatives advised in the treatment of this disorder, chloral hydrate promises the best, and should be administered as is advised in strychnine poisoning. This drug may prove ineffectual; if so the tincture of the aconite root should be given by injection; three drops every two hours until relief is obtained. If the result from the use of aconite is discouraging, one eighth of a grain of morphine may be administered subcutaneously, and repeated every three or four hours.

Of the utmost importance is abundant nourishment and stimulation. A cup of milk or beef tea with a raw egg and a tablespoonful of brandy should be given by the rectum every three or four hours, unless it can be taken by the mouth. It is to be remembered that after the spinal cord becomes affected it acts independently, and its irritation does not subside on the removal of the exciting cause.

CHAPTER XV.

GENERAL DISEASES.

DISTEMPER.

Distemper is a fever originating through infection of the system with a certain peculiar poisonous matter, the nature of which is unknown. By some it is denominated the specific, catarrhal fever of the dog; the name being suggested by the inflammation of the mucous structure which characterizes the affection.

Causation. — When contagion plays so important a part, the existence of a specific germ must be admitted for a disease so clearly defined and so well characterized. Accepting this theory we reach the natural conclusion that distemper never originates spontaneously, but is dependent upon a transmission, a continued propagation of the disease poison. Some writers found sufficient explanation for the origin of the affection in exposure to damp and cold, ill ventilated kennels, poor food, improper feeding and the like, but as ideas have gradually developed and important advances have been made in tracing to fixed conditions of truth the origins of certain diseases, once believed to have originated by equivocal generation, we can scarcely doubt the doctrine of continuous propagation, and can no longer realize how every considerable deviation from a proper observance of the laws of health should be followed by the production of a particular disease.

That influences which tend to debilitate are potent factors in the extension of the disease is not questioned, but that it originates only when the specific germ of the disease exists by itself, or has been introduced, must be believed.

The poison of distemper can be conveyed from one dog to another by contact; but it is true that direct contact is not a necessary condition for transmission; it can be conveyed by the air and in many other ways.

Hunting in his admirable treatise on this disease states: — "There are certain circumstances favorable to its origin — the congregation together of large numbers of dogs, the transit of dogs through the same places and in the same vehicles, contact with strange dogs—all of which are simply facilities for the conveyance and transmission of the specific poison of the disease. In kennels of hounds the most common time for outbreaks of distemper to appear is when the young entries arrive, *i. e.*, when the young hounds which have been farmed out at various places, return home. Any of these may have been in contact with a dog suffering from distemper, and if one be infected all in contact with it suffer. The possibilities of the introduction of disease are in exact proportion to the number of places from which the animals are collected. Dog shows are a fruitful cause of the spread of distemper. No matter how well managed they may be, the mere fact of collecting from various places a vast number of dogs, renders a large show almost certainly a focus from which the disease is disseminated widely. In some cases a large show has been the means of infecting nearly every puppy sent to it. The malady is not at once made evident, but shows itself a few days after the return of the animals. This cannot altogether be prevented. Dogs just convalescent and free from any apparent disease are sent to shows, and are capable of communicating distemper to others. Even healthy dogs having an immunity from the disorder, because of a previous attack, may carry the disease in their coats if they have been in recent contact with a diseased dog, and so spread the malady without ever being suspected. Some of these risks might be obviated by requiring all exhibiters to state that for one month previously their kennels have been free from contagious disease, and that their dogs have not been in contact with diseased dogs for a week anterior to the show. Railway boxes, hampers and portable kennels are sources of possible contagion which might be lessened were disinfection and washing more methodically carried out. Distemper can be easily transmitted from a diseased to a healthy dog by a nasal discharge. It is often spread by means of the food which a diseased dog has left. Its contagiousness probably no one disputes, but most men believe that there are cases which rise independently of any contagion. The basis of such a theory is the fact that in some cases the exact method of communication is not traceable. Until a case can be found under circumstances which render the conveyance of the specific poison an impossibility, this theory has absolutely no facts to support it. The method of communication of disease is often difficult to trace; we can satisfactorily account for ninety cases out of a hundred by contagion, and it can hardly be called begging the question to say that the odd ten, presenting the same sign, running the same course, and being equally contagious, are due to the same cause that actuated the ninety."

In accepting the theory of the existence of a specific poison, and that distemper can only be communicated by contagion, the alleged causative conditions, such as exposure, debility, improper feeding, ill ventilated kennels etc., are to be regarded no longer as exciting causes, and are to be considered only in so far as they may increase susceptibility, favor extension, and possibly determine the severity of the disease.

It is reasonable to suppose that inasmuch as the general system becomes weakened by ante-hyenic influences the power to resist contagion is lessened, the severity of disease is in a measure intensified, reaction and repair are retarded, and mortality is thereby increased.

The poison of distemper, as in other contagious diseases, can reproduce itself under favorable conditions to an endless degree; it retains its vitality and power of infection for a long time outside of the organism which produced it; it has no special stage of development to pass through on the way from the affected organism to the one to be affected; but at the time of infection it is essentially in the same condition as when given up by the organism yielding it. That a simple attack of distemper successfully overcome, imparts an immunity from it for the remainder of life, is a rule with only occasional exceptions. In explanation of the theory of immunity bestowed on man by certain contagious diseases, it has been suggested that something analagous to the ferment process takes place. If yeast is placed in a fluid containing sugar, fermentation follows; when all the sugar has been destroyed, and fermentation is complete, it cannot be produced again by a further addition of sugar; the fluid responds no longer to the action of the yeast. It is conjectural that during certain contagious diseases there is a destruction or change in the body of some chemical or constituent, the presence of which is a necessary condition for the retention or development of the disease. It is not unreasonable to suppose that in distemper something analogous takes place.

The disease under consideration has been compared by some to the typhoid fever of man; that it differs in its important symptoms, and appearances as observed after death is true, but it seems that the greatest disparity of all has been overlooked; typhoid fever belongs to the class of miasmatic contagious diseases and is propagated in an essentially different manner.

Distemper can be conveyed by contact from a diseased animal to a healthy one and produce the disease in him, while typhoid fever is not contagious in the proper sense of the word, for it is never transmitted by direct contact. On the profound difference in the mode of propagation depends the essential difference in the two diseases. A far greater analogy exists between typhus fever of man and distemper; it belongs to the same class of contagious diseases; the specific germ is evidently transmitted and propagated in much the same manner. The incubation period, the febrile stage, the duration, the self limitation, and many essential symp-

toms are alike peculiar to both. Although distemper attacks dogs of all
ages it is more common in early life, the disease in old dogs being rarely
observed, for the reason that nearly all have experienced an attack and
are absolutely, or in a great measure insusceptible to the poison.

Symptoms. — The period of incubation, or slow development after exposure to contagion and preceding the first symptoms of distemper, is from
four to fourteen days; probably the latter is the limit in the large majority
of cases. An attack is usually preceded by dulness, disinclination to
exertion, loss of appetite, and chilly sensations as shown by shiverings.
Then rapidly follow the symptoms of a common cold, sneezing, dry,
husky cough, and a watery discharge from the eyes and nose. The fever
soon becomes prominent, the pulse full and decided, the nose hot and dry,
and the lining membrane of the eyelids reddened, the tongue slightly
coated, and the secretion of saliva diminished. Vomiting is not uncommon, food is rejected or eaten sparingly; there is thirst, a tendency to
constipation, and still greater prostration. On the second or third day the
discharge from the eyes and nose becomes purulent, gluing the inflamed
eyelids together, and drying in crusts around the nostrils as the fever
grows more intense. Muscular weakness and prostration increase, the
animal is less easily disturbed, and when in motion hangs his head, makes
but little effort, soon lies down, becomes indifferent, or dozes into an
unquiet sleep.

The cough at first short and dry, loses its husky tone as the mucous
secretion becomes abundant, and in mild cases soon ceases. As the disease progresses the pulse and respiration are increased in frequency; the
appetite more often entirely disappears; the stomach and intestines become irritable, and a stale, even offensive odor proceeds from the mouth.

The fever runs a course of variable intensity; in uncomplicated cases
it usually reaches its height in four or five days, and then rapidly subsides.
So favorable a result is by no means the rule; more often complications
arise, prolonging the illness, and rendering recovery more uncertain. If
reaction is delayed, and the constitutional disturbances persist, prostration becomes more marked, and is accompanied with rapidly increasing
emaciation.

In the progress of distemper seemingly trifling influences often induce
disastrous results; a system vitiated by the specific poison of the disease,
deranged and prostrated by an intense fever, its vital energies exhausted,
has but little resistive power, and is predisposed to further disorder. The
digestive organs already suffering from catarrhal conditions, are easily
inflamed by medicine or food unwisely selected; an irritable condition of
the mucous membrane of the air passages renders it easy to excite a
catarrh of the smaller bronchial tubes; weakness of the heart's action

invites certain congestive changes in the lungs; pneumonia constitutes a frequent and serious complication of distemper. In this disease as in all others characterized by prolonged high fever, certain degenerative changes occur in the internal organs, the vessels, the blood, and in the muscular system. The change in the liver is shown by the markedly diminished secretion of bile, the constituents of it being retained in the blood and giving rise to jaundice.

The muscular degeneration is a part of the manifestation of the fever, and depends upon the derangement of a center regulating the nutrition of the muscles. In many cases of distemper the functions of the brain become more or less disturbed and weakened; occasionally effusions take place within the cranial cavity; epileptiform or general convulsions, meningitic symptoms, apoplectic seizures are in a high degree prejudicial to the chances of recovery; for they indicate some unusual cause of disturbance, often a grave disease of the brain or its attachments. There is a decided tendency to constipation among some, while among others diarrhœa and even dysentery exist.

During the period of dentition puppies are predisposed to convulsions and intestinal catarrh; the transition of the latter into a more serious disease of the bowels, attended by profuse discharges and rapid emaciation, occasionally occurs, and in some instances terminates in collapse and death. The influence of age is strikingly manifested in distemper; the fever usually running a course of greater intensity among puppies. The presence of worms is a complication which by functional disturbances prejudices recovery.

Of notable influence upon the course of the disease under consideration is the constitution of the dog attacked. Some breeds are more sensitive and excitable; others in comparison might be called sluggish; in the former the general disturbances of the system are more pronounced, and brain symptoms are more easily induced. It is presumed that the in-and-in bred animals for this reason have less resistant capability; the same may be said of dogs improperly fed and generally neglected.

Occasionally distemper is aggravated and protracted in its course by irritable and painful ulcers of the cornea, resulting from an inflammation of the eyes, at first superficial and unimportant. An eruption of the skin, and loss of hair from some portions of the body, is a frequent complication, due doubtless to weakened vitality and impaired nutrition.

Thus it will be seen that during an attack of distemper the predisposing influences which invite further disorder are many, powerful, and not infrequently beyond control. The duration of the disease depends upon the complications which may arise; where none appear the fever reaches its height about the fifth day, then gradually declines, and recovery is complete at the end of the third week. In cases running a fatal course death usually occurs before the fifteenth day. Chorea and paralysis are two important affections consequent upon certain attacks of distemper; the latter usually affecting the hind quarters only.

The duration of complicated cases of distemper is problematical. In many instances convalescence is for a long time delayed, and recovery takes place slowly and is only completed after many weeks. When we consider that the important organs in all parts of the body are weakened and otherwise injured by the distemper process, that important constituents have been destroyed, that nutrition is retarded by defective digestion, it becomes intelligible to us that complete restoration is slow after the severer attacks of the disease.

Treatment. —Distemper is a disease which tends intrinsically to end after a certain time, and is therefore self limited. A fatal result is rarely due to the intensity of the disease; death is generally attributable to complications. These facts are of considerable importance in estimating the amount of curative influence exerted by methods of treatment. No reliable means are yet known which can be depended upon to arrest this disease, or even shorten the duration of the febrile career. Active measures employed simply because the distemper exists, without reference to events connected with it, are contra-indicated. The progress should be carefully watched and appropriate treatment be employed as unfavorable symptoms arise; attempts should not be made to abridge or arrest the disease, but rather is nature to be assisted, and remedies be employed to aid in conducting it to a favorable termination.

The management of distemper involves not only the exercise of judgment in the employment of medicinal remedies, but of attention to sanitary and sustaining or supporting measures; a most important object being to obviate the tendency to death by exhaustion, and to forestall a degree of prostration dangerous to life.

When the disease makes its appearance the affected animal should be isolated from others and placed in dry, well ventilated quarters. The importance of an abundance of pure air cannot be exaggerated; in cold weather moderate artificial heat will be needed. The presence of disinfectants about the room are advised, and preparations of lime and carbolic acid in powder are to be preferred.

It is an urgent necessity that as complete bodily rest as possible be secured from the very beginning of an attack; restraint must therefore be enforced until the period of danger has passed. To unload the bowels at the onset is important, and the syrup of buckthorn and castor oil, of each two or three teaspoonfuls, may be given.

Appreciating the disturbing influence of a high fever it is our first duty to endeavor as far as possible to hold this under control, and for that purpose the following may be employed for the first two or three days.—

℞ Tr. Aconiti Rad.   gtt. xxx
   Potass. Chloratis   ℨ i
   Ammon..Mur.   ℨ ss
   Spts. Æth. Nit.   ℥ l
   Aquæ   ℥ ij

Ft. Mist. Sig. Dose one teaspoonful.

While the fever is intense this mixture may be given in teaspoonful doses every two hours, and discontinued as soon as the high temperature abates.

Taken in the first stage of the disease, the administration of quinine in one large dose generally results in materially reducing the fever. If employed, eight grains should be given in pill form; this is not to be repeated, but the fever mixture can be subsequently used if needed. A constant supply of pure drinking water is an essential, and the chlorate of potassa should be added to it, in proportion of two teaspoonfuls to a quart.

It is a matter of the greatest consequence that an affected dog be well nourished, and a proper amount of food should be insisted upon, even if no desire for it exists. During the first day or two he will doubtlessly eat sparingly of broths, beef tea, or milk, and what is taken voluntarily will suffice. The appetite will then more often nearly if not entirely disappear; when all else is rejected, raw beef in many cases will still be acceptable, and should be allowed. To render it more easy of digestion it should be scraped, and one or two tablespoonfuls given every two hours. The addition of gelatine and raw eggs to the beef renders it still more nutritious. In some instances no nourishment of any description will be taken voluntarily; force must now be employed, measured by the manifest need to support and sustain the vital powers.

The irritability of the stomach and intestines, and the tendency to vomiting and diarrhœa must not be overlooked, and the most easily digested and concentrated nourishment be selected. Milk and lime water, strong beef tea, and beef essences are indicated. Food should be poured down the animal's throat at least four times a day; in shorter intervals if emaciation is great and rapidly progressive. No more than a cup full should be administered at one time. If vomiting is excited, the quantity of nourishment must be divided and given oftener. Milk and lime water is easily borne on an irritable stomach, but if possible it should be alternated with beef tea and beef extracts; the addition of a raw egg to each cup full renders the nourishment more strengthening and no more difficult of digestion. After the fourth or fifth day it will be well to give with the food, four times a day, one grain of quinine in pill form.

Stimulants are never to be withheld until that point is reached when prostration indicates a failure of vital powers imminent. If the animal seems fairly well sustained, and the nourishment is well borne on the stomach, a teaspoonful of sherry wine may properly be added to each cup full of milk or broth; as the disease progresses and he grows weaker, the amount should be increased to a tablespoonful; if failure is still progressive, then brandy must displace the wine and should be given in quantities from a teaspoonful to a tablespoonful as the need is urgent. Even brandy may prove insufficient to stay exhaustion; then to it must be added one half a teaspoonful of aromatic spirit of ammonia.

It is presumed that in an uncomplicated and properly treated case of distemper, the need of excessive stimulation will rarely occur; cases where neglect and injudicious dosing have induced extreme depression will be more often seen, and not infrequently will complications exist without being at once recognized. The general disturbance will however be apparent and demand the use of stimulants as advised. Treat the affected animal and not the disease is a wise rule to follow.

It is not always easy to determine just when pneumonia and other affections become complicated with distemper, and to remain inactive while the animal is growing worse, until positive evidence of their existence is manifested beyond a doubt, would be extremely hazardous; the study and treatment of symptoms is therefore imperative.

Vomiting may possibly occur notwithstanding every care in diet, and demands the use of subnitrate of bismuth in five grain doses, four times a day. When the stomach is too irritable to retain milk, broths, beef tea, etc., they must be withheld for a few hours, and only scraped raw beef pressed into a pill form be given with the bismuth.

Constipation if it exists will rarely if ever need medicinal treatment; liver and oatmeal have a decidedly laxative action, and if they prove insufficient a rectal injection can be administered.

Pulmonary complications are largely attributable to exposure to cold, and may appear at any time during the course of the disease; more often they occur after the height has been reached, and during the period of convalescence.

Diarrhœa is the more often due to dietetic errors, and the cause should be obviated. Two or three discharges from the bowels daily call for no remedial agents; in more relaxed conditions, paregoric in teaspoonful doses alone or combined with twice that quantity of chalk mixture, and repeated as needed, should be administered.

The discharge from the nose and eyes should never be allowed to accumulate, a sponge and borax water being used in its removal. During convalescence exercise should be restricted, and exposure carefully guarded against. Changes in diet should be made cautiously, measured by the powers of digestion. If recovery is slow tonics are indicated, cod liver oil will prove especially valuable. The different complications will be found elsewhere considered as distinct diseases.

DIPHTHERIA.

While diphtheria is an affection which very rarely attacks dogs, the fact that it has been observed among them renders a brief description warrantable.

It is a disease which invades especially the membrane of the air passages, and general infection follows the local expression. The local disease appears as an inflammation of the mucous membrane, especially of the mouth, pharynx, nose, larynx, and deeper air passages, and precedes the formation of a greyish-white membranous exudation.

The constitutional affection has the character of infectious diseases, and is essentially the same as in all dependent upon blood poisoning. The problem is still unsolved whether diphtheria is at first a general disease, and poisoning of the blood and the local affections of the mucous membrane are secondary manifestations, or whether the disease begins by local infection, from which point the poison is taken into the system and the whole organization becomes contaminated.

It is generally accepted that the germ of the disease first fixes itself upon a certain part of the body, more often the mucous membrane, and there at its point of attachment it excites a local affection, then through absorption the poison penetrates the tissues and is soon carried throughout the whole body.

Diphtheria is an epidemic disease, and like all others has an especial cause, a contagious virus or infectious miasm. Certain conditions favor the propagation of the disease; impure air and poisonous exhalations from decomposing filth are important factors in its production. Among the reported cases of the disease among dogs, it is observed that the onset in some was marked by great febrile disturbance; in others the constitutional symptoms were less severe. In the larger proportion the disease attacked the mucous membrane of the throat, and more rarely were the nasal passages first involved. In the former the throat presented a dark red, swollen, glistening, appearance; the glands of the neck being enlarged and tender, and swallowing difficult. In some instances ulcers were seen on the tonsils.

When the disease assumed a concealed or nasal form, it manifested itself by general depression, fever of a low type, and a thin, reddish, offensive odor from the nose. The glands of the neck were much swollen in these as in the other cases. Where the disease has appeared it has generally proved fatal. When recovery took place, in some loss of sight, in others paralysis and chorea followed.

Cases are on record where diphtheria has been without doubt transmitted from man to dogs; this fact, strongly evincing that the disease is analogous in both, is important in considering treatment.

If the deeper air passages become involved as shown by harsh, croupy, spasmodic breathing, the case may be considered hopeless. If the disease takes on the concealed or nasal type the nose must be frequently syringed with lime water or carbolic acid solution; two grains to an ounce of water. An application of dry powdered sulphur, blown into the throat through a tube or by means of a small bellows, is advised. This operation should be performed every half hour if possible. For internal treatment referable to the congestion of the mucous membrane, chlorate of potassa should be given. An ounce may be added to a pint of water, and of this solution two or three teaspooufuls can be given every hour. Beef tea, raw eggs, milk, and brandy should be forced into the animal at frequent intervals, to sustain the vital powers.

RHEUMATISM.

Rheumatism not infrequently attacks dogs and manifests itself in much the same manner in them as in man. The disease may take on an acute or chronic form; the same morbid principle is supposed to give rise to both, but in the latter it is present in a lesser degree.

Causation. — The causes of rheumatism have never been fully determined. It has been commonly attributed to exposure to cold and damp; influenced largely by the seasons of the year, etc. That these agencies alone are incapable of producing the disease is generally accepted. A morbid peculiarity of the constitution, a special predisposition seems to be requisite for the causation. When this susceptibility to the disease exists then exposure doubtless acts as an exciting cause.

Symptoms. — In acute rheumatism fever is always present, the skin is hot, the appetite lessened, the thirst great, the pulse rapid, the animal restless, movements difficult and painful. The tongue is generally coated, the breath offensive, the respiration quickened, and constipation more often exists.

The essentiality of the disease consists in a swelling of the various joints. Touching, and still more the movements of the affected parts is extremely painful, and the animal with an expression of the utmost anxiety, will guard over and seem to protest against an examination. The affected parts are hot and swollen. Rarely are several joints attacked at the same time; usually the disease partially runs its course in one before another is involved.

When recovery takes place the swellings of the joints disappear without leaving any traces of the disease behind them; in some instances a slight weakness and painfulness will for a time linger as an evidence of the previous existence of the malady.

Treatment. — To place an animal suffering from the disease in warm, dry quarters, is the first important step to take. The painful joints should be enveloped in cotton batting; the bowels opened by the syrup of buckthorn; a bland unstimulating diet allowed, and the following given. —

R Sodii Salicylatis   ℥ ij
      Aquæ   ℥ iij
Ft. Mist. Sig. Dose one teaspoonful.

This mixture should be administered every two hours, in a little milk, until improvement is observed, then at longer intervals to complete recovery. The tendency of the disease to relapse should not be forgotten, and during the convalescent stage exposure should be carefully guarded against.

Chronic rheumatism rarely follows as a result of the acute form. The general symptoms are stiffness, some pain of no great intensity, tenderness comparatively slight, swelling of the joints, little or no fever. The constitutional disturbance is trifling if any, the appetite remains good, the different functions of the body are nearly normal, and nutrition is unaffected. Radical changes in the weather are generally followed by more acute symptoms in the affected joints.

Local treatment in chronic rheumatism is important, and benefit may be anticipated from the use of the tincture of iodine or blistering the joints involved. Measures having reference to improving the general health and the tone of the system are essential, and embrace tonic remedies, nutritious diet, exercise, etc.

The iodide of potassium is a valuable remedy in many cases, and may be given in three or four grain doses, two or three times daily. If an iron tonic is indicated the syrup of the iodide can be substituted; dose fifteen drops three times a day with the food.

LUMBAGO.

This affection is a form of the so-called muscular rheumatism, and differs essentially from that disease which attacks the joints. The muscles of the loins are the seat of the disturbance, and the animal's movements are stiff and painful; he stands with back arched, and when he walks his limbs are dragged as though partially paralyzed. In certain positions he suffers but little excepting an occasional cramp-like twinge, which is sometimes very severe and elicits a sharp outcry. While the parts affected are sensitive, firm pressure will be borne and seem to afford relief.

There are few or no constitutional symptoms, slight if any fever, and the appetite is rarely impaired. Aside from the tenderness there are no local signs, such as heat, swelling, and redness. In chest-founder or kennel-lameness the affection exists in the muscles of the shoulders and fore legs. The disease may appear either in the acute or chronic form; the duration of the former varies from a few hours to several weeks. The chronic form is obstinate, and persists usually for months. Exposure to cold and damp is an important factor in the causation of this form of rheumatism.

The local treatment consists of soothing applications in the acute form, and blisters or stimulating liniments in the chronic. Measures to improve the general health are important, and iron and other tonics are to be given if the condition of the animal suggests their need.

ERUPTIVE FEVERS.

Cases of small pox and measles occuring among dogs are on record, and the symptoms of the former malady are given at length by some authors. In the one reported case of measles it was stated: "A dog licked the hand of a child lying in bed, and on whom the measles eruption was at its height. Twelve days later the dog sickened and suffered for two days with nasal discharge; and four days later died, with marked congestion of the throat and air-passages."

Mr. Fleming in the Veterinary Science thus discusses small pox: "This is a rare malady, and may be developed directly or by contagion; it is supposed to be also produced by the variola of man and of the sheep. It chiefly affects young dogs, although old animals are not exempt. One attack ensures immunity for the remainder of the dog's life.

"*Symptoms.*—The disease commences with fever, which continues for two or three days, and is followed by the appearance—over a large surface of the body, though rarely on the back and sides of the trunk—of red points, resembling flea-bites, which are quickly transformed into nodules, and then into vesicles. The contents of these become purulent, and finally dry into a crust, whose shedding leaves a naked cicatrix.

"In the dog, as in the sheep and pig, there are different forms of the disease, and it is benignant or malignant accordingly. Puppies nearly always succumb, and, on a necroscopical examination, it is not unusual to find various pustules on the mucous membrane of the respiratory and digestive organs.

" *Sanitary Measures.* — The disease being contagious, though the virus does not appear to be very volatile, it is necessary to isolate the sick, and take due precautions that the contagion is not carried from them to healthy animals.

" *Curative Measures.* — Careful dieting, a dry and moderately warm dwelling, cleanliness, and abundance of fresh air are the essentials in the curative treatment.

"An emetic in the early stage of the malady has been recommended as likely to be useful. Afterward the treatment must be purely symptomatic."

DROPSY.

Dropsy, a morbid serous transudation into any of the cavities, is never a primary affection but only a symptom, a sequel of many chronic diseases, particularly those of the liver. Dropsies are not to be confounded with effusions; the former is not the result of inflammation, and the morbid condition on which it depends is situated elsewhere; the serous membrane in which it occurs being free from disease. Liquid effusions are exudations involving generally if not always inflammation.

Dropsies receive different names according to their situations; when located in serous cavities, they are designated by prefixing hydro to the name of the membrane. Dropsy within the head is called hydrocephalus; hydrothorax, when within the chest; ascites, when within the abdominal cavity. The latter is the most frequent form seen in dogs, and deserves especial consideration; the others being of less practical importance.

Symptoms. — The enlargement first directs attention to the abdomen, which becomes equably large and fluctuating, not filling at one part more than another; the accumulation of liquid taking place without pain or tenderness. The fluid usually forms rapidly, and the distension soon occasions disturbances of the internal organs by compression. The appetite becomes impaired, vomiting is an occasional symptom, the urine is scanty, respiration is embarassed, the pulse more rapid and feeble; constipation may alternate with diarrhœa, and pressure on the veins give rise to dropsy of the lower limbs. In the progress of the affection emaciation becomes marked, the lips, tongue, and gums palid, the pulse thin and thready, the breathing more hurried, threatening suffocation; the animal can no longer lie down; death finally occurs from exhaustion or obstructed respiration.

Diagnosis. — The diagnose of ascites can generally be made without

difficulty. Blaine says: "Dropsy of the belly may be distinguished from
fat by the particular tumor that the belly forms, which in dropsy hangs
down, while the backbone sticks up, and the hips appear prominent
through the skin; 'the hair stares also, and the coat is peculiarly harsh. It
may be distinguished from being in pup by the teats, which always enlarge
as the belly enlarges in pregnancy; but more particularly it may be distinguished by the undulation of the water in the belly, whereas in pregnancy there is no undulation. The impregnated belly, however full, has
not that high tense feel nor shining appearance observed in dropsy. There
may be also inequalities distinguished in it, which are the puppies, and,
when the pregnancy is at all advanced, the young may be felt to move.
The most unequivocal mode however, of detecting the presence of water
is by the touch. If the right hand is laid on one side of the belly, and with
the left hand the other side is tapped, an undulating motion will be per-
ceived, exactly similar to what would be felt by placing one hand on a
bladder of water, and striking it with the other."

Prognosis.—The diseased conditions on which dropsy is dependant
are generally incurable; under treatment improvement for a time in many
cases occurs and the ascites disappears but soon returns, and death ultimately results.

Treatment.—The purpose of treatment is to remove the dropsy, and
improve the morbid conditions on which it depends. To accomplish the
first the diet should be milk only, and cathartics must be relied upon.
One-sixteenth of a grain of Clutterbuck's elaterium should be given every
two or three hours, until profuse liquid discharges from the bowels are
produced. Afterwards the doses are to be repeated according to circumstances. The strength of the animal must determine how far the cathartics
should be pushed.

As the dropsy disappears and symptoms of suffocation are no longer
imminent, the tincture of the chloride of iron should be given in fifteen
drop doses, well diluted, three or four times a day. A more generous diet
can then be allowed, and other measures, calculated to improve the general
strength and add tone to the system, should be employed.

ANÆMIA.

Anæmia, or poverty of the blood, is a morbid condition occasionally
observed among dogs. The best illustration of the affection is afforded by
profuse hemorrhages. The causes are obvious in some cases; in others
they are not assignable. An excess of starch and insufficient fleshy food,
prolonged suckling of young, too frequent whelping, deranged digestion,

damp, ill-ventilated kennels, and insufficient exercise are among the causes productive of anæmia. In a large proportion of cases the affection is associated with and depends upon other diseases which involve an expenditure of blood constituents.

Symptoms. — In animals affected with anæmia the mucous membranes of the mouth, gums, and lips are paled, the tongue is white and dry. the temperature lessened, the action of the heart feeble. the pulse small and weak, the muscular strength is diminished and occasionally there is but little power in the limbs. Slight exertion induces great fatigue; the manner is languid and spiritless; the secretions from the bowels and kidneys are scanty. There is a deficiency of functional energy of all the different organs of the body.

Treatment — The first objects are to ascertain and remove if possible the cause or causes upon which anæmia depends. Associated diseases call for especial treatment elsewhere considered. To restore the normal constituents of the blood, a nutritious diet consisting largely of meat should be allowed; to stimulate the appetite and improve digestion by bitter tonics, such as quinine, is advisable. Iron is a special remedy in this disease, and the tincture should be given in fifteen drop doses, twice or three times daily with the food. Pure air and exercise contribute largely to a cure, by increasing the energy of the organs of the body and renewing the general strength.

PLETHORA.

In general terms the morbid effects of plethora are the reverse of those due to anæmia. The causes are overfeeding and insufficient exercise.

Symptoms. — A plethoric animal has a full, bloated, congested appearance; the mucous membranes are reddened; the heart's action is more powerful, the pulse being full and strong; a feverish tendency exists, and a dull, heavy, sluggish manner is characteristic. In this condition the brain is more easily excited and congested, involving a liability to convulsions.

Treatment. — Plain unstimulating food, reasonable in quantity, enforced exercise, and the occasional use of laxatives.

OBESITY.

Dogs not infrequently become fattened to a morbid degree, and whenever it exceeds the limits of health it may appropriately claim treatment. Insufficient exercise, over-feeding, and an excess of saccharine and starchy

food are the causes assignable. In some breeds a constitutional fat producing tendency seems to exist.

Treatment. —Enforced regular exercise; a deprivation of sweets, vegetables, and starchy food; raw, lean beef, in a reasonable quantity should
be the principle article of diet.

ℛICKETS.

Rickets is a peculiar unhealthy condition of the whole body. This disorder is quite common among young dogs, and more especially the larger
breeds.

Causation. —Debilitated or rachitic sire or dam may transmit a taint
to their offspring, or it may arise from improper feeding, lack of sunlight,
impure air, close confinement, and impaired nutrition. In some instances it is developed rapidly after other diseases, which have left the system
in a state of debility.

Symptoms. —To correctly comprehend the changes in rickets it is necessary to briefly consider the physiological growth of bone. Every tubular bone grows in length and thickness; in length by the deposit of new
layers of cartilage-cells, in which limy salts are then deposited. It grows
in thickness by the addition of new layers of bony substances immediately beneath the covering of the bone. As the growth in thickness is
much more insignificant, and progresses slower than in length, the disturbances of the physiological growth at the cartilaginous ends are more
marked and liable to occur. While the bone is enlarging externally by
new formation, absorption takes place in the center. In rickets this absorption goes on, but prompt ossification is suspended or but imperfectly
performed.

No bone affected by this disease retains its normal form. The angles
become rounded; the long or tubular bones cease to grow in length; the
ends swell and become bulbous. The weight of the body causes the bones
of the fore-legs especially to bend, and permanent angular deformity
results. The bones of the head seem to grow at the expense of the body,
and it appears swollen and ill-shaped.

The coat is rough and staring; the gums and lips pallid; the muscles
flabby; the general appearance of a rachitic animal is unhealthy, stunted,
and unsightly. Digestion may remain undisturbed, but occasionally diarrhœa supervenes; then the disease becomes materially aggravated.

Treatment. —Pure fresh air, sunlight, and dry, well ventilated kennels
are absolutely necessary. Lime water should be given freely with the
food, which ought to be generous and nutritious. Cod liver oil is the one

important medicinal remedy to depend upon, although iron and quinine can be added to the treatment if their need is indicated.

In large breeds there is a liability to deformity which might be easily mistaken for rickets. It results from confinement and over-feeding, the body becoming too large, and too heavy for the legs to sustain. In such cases the quantity of food should be lessened, and by regulated exercise the weaker parts strengthened.

SCROFULA.

Scrofula is a constitutional and specific malady, involving the glandular system; a general debility, with a tendency to indolent inflammatory and ulcerative diseases. Its occurrence is not uncommon among dogs.

Causation. — Scrofula may be hereditary; if not it can be created by influences capable of lowering the vital energies; by improper food or insufficient of it; by general neglect, lack of exercise, cold, damp kennels, and want of pure air and sunlight.

The disease may appear after other diseases have for a time existed and depressed the general health and impoverished the system.

Symptoms. — When the scrofulous constitution exists, the functions are but irregularly performed. Digestion is impaired; the appetite is capricious; the mucous membranes pale and flabby; the bowels constipated; the coat dry and staring, and the abdomen distended.

In certain cases the disease may exist, remain hidden as it were, awaiting some unusual cause to excite its development; puppies so affected grow rapidly, and show in outward form but little evidence of the taint. In other cases the growth is stunted, the joints large, legs bent, head ill-shapen, and the animal is in general appearance an unsightly object. As the symptoms of scrofula become prominent emaciation is usually rapid, the muscles weak and flabby.

Treatment. — The indications are first to correct abuses. The food should be nutritious and abundant, consisting largely of meat. Free exercise in the open air and sunlight, and dry, well ventilated kennels are indispensable. Perfect cleanliness and frequent bathing are of great importance.

Among the medicinal remedies cod liver oil takes precedence. If the appetite needs stimulating quinine is indicated. When the mucous membranes by pallor show a poverty of the blood, iron should be given for a long time.

Animals which have ever exhibited the symptoms of a scrofulous taint, should never be used for breeding purposes.

MARASMUS.

The meaning of the word marasmus is atrophy; wasting of the flesh; emaciation. It is properly a disease of the mesenteric glands and the follicles of the intestines; the latter inflame, adhere together, and ulcerate.

Symptoms. — Emaciation, nose dry and hot, abdomen swollen and hard, eyes reddened and watery, skin dry and harsh, coat staring, the breath foul, thirst great, and appetite voracious. Bowels at times constipated, the discharges clay-colored and offensive; occasionally diarrhœa exists.

Treatment. — The diet should consist largely of meat given in small quantities and at short intervals.

Cod liver oil and iron are indicated, and should be given cautiously at first.

Pure air, sunlight, and exercise are important factors to aid in restoring the integrity of the system.

CHAPTER XVI.

SURGICAL AFFECTIONS.

WOUNDS.

A wound is a separation of continuous parts by violence. It is of practical importance to consider four varieties:—incised, punctured, lacerated, contused.

The incised wounds, made by clean-cutting instruments, are usually produced with the least violence, and generally the most easily repaired. The punctured, those made by instruments of greater length than breadth, including stabs etc., are dangerous from their depth, and from the possibility of hemorrhage and injury to the internal organs; abscesses are liable to follow deep wounds. In the lacerated, the parts are torn, and the contused are effected by bruising; these wounds are commonly produced by greater violence, heal more slowly and invite the formation of pus; they generally bleed less than the incised or punctured, because arteries when torn, contract and close more than when cut.

Treatment.—The indications in the treatment of wounds are to check bleeding; remove foreign bodies; bring the divided parts together, and secure them in a natural position; to promote healing. Pressure and the application of cold will generally suffice to control hemorrhage, unless an artery of considerable size is wounded, in which case if it persists in bleeding, a surgeon should be called to ligate.

Foreign bodies can be removed by the fingers or forceps, and dirt, gravel, etc. washed away. If the wound gaps open the edges should be brought together and held by sutures, for which silk is preferred.

When a dog has been wounded and stitching is necessary, one must not shrink from the duty through fear of violence; rarely will any be encountered, for the noble brutes possessed of a human intelligence are conscious that a kind hand inflicts the pain to repair their injury, and they undergo the operation with a fortitude which excites wonder and admiration. In all extensive wounds stitches should be introduced with an interval of a half an inch between them, entering the skin well back from the edges to

prevent their tearing out. After the operation several thicknesses of antiseptic gauze, if that can be obtained, should be bound over the injured
parts, and left undisturbed for three or four days, at the end of which time
the stitches can be removed and the wound again dressed as before. When
the antiseptic gauze is unattainable, a compress of linen should be applied and kept damp with a solution of carbolic acid, two drachms to a
pint of water.

In the treatment of extensive and deep punctured wounds as in stabs, it
is to be remembered that the cut in the skin is liable to unite, and thus by
confining the discharge give rise to deep-seated abscesses. To prevent
this a small drainage tube can be introduced, and left in the wound for
two or three days and then removed permanently, or after being cleaned
and disinfected be again returned, and shortened from time to time as
the healing from the bottom progresses. A dressing of the antiseptic
gauze should be kept lightly and constantly applied. When the proper
tubing cannot be secured, a cord made of several strands of ligature silk
can be substituted. In smaller punctured wounds it is wise to bandage
them with some degree of firmness, and keep the parts at rest as much as
possible; then if pain, heat, and swelling follow, an abscess has resulted;
an incision should be made and poultices applied.

Lacerated wounds are to be dressed antiseptically, and treated much
the same as incised wounds after being stitched. Those from bites should
be first freely cauterized.

Contused wounds if superficial and deserving of treatment, generally
do well under cold applications. When the deeper tissues are severely
injured and the vitality of portions are destroyed and must separate
from the healthy flesh, poultices should be applied to hasten the sloughing process and be persisted in until the wound is clean and healthy, after
which it can be treated as an ulcer, the oxide of zinc ointment being
appropriate Old, indolent, suppurating wounds which heal slowly need
stimulating occasionally; for which purpose caustic can be used, not extensively but simply touched here and there over the entire sore.

When deep wounds or abscesses have not been thoroughly healed from
the bottom through a defect in bandaging, or in providing proper outlets
for the discharge, a fistula results. This is a narrow canal opening on the
surface and leading to the suppurating cavity.

The treatment is to enlarge the opening; keep it open and syringe out
the cavity with a strong solution of the nitrate of silver, grains thirty to
water an ounce.

UMBILICAL HERNIA.

Umbilical hernia is a rupture, or unnatural protrusion of the viscera at the naval. The affection is more common among puppies, but may appear at any period of life. When occurring in older dogs it is more often met with in bitches, resulting from large litters and frequent pregnancies.

Various methods of treatment have been suggested, but bandages and other attachments cannot be depended upon, and a surgical operation alone promises certain success.

Mr. Lowe has proposed a method to produce a radical cure of this affection, which is commended as simple, comparatively safe, and easy to perform. He describes the operation as follows :—

"The instruments absolutely needed are a thin, clean, sharp knife, a curved surgeon's needle and silk; also, a little iodoform, and some carbolized water. The two latter act as antiseptics, and fill the places of the carbolic spray fairly well. After the patient has been thoroughly etherized, wet the fingers, knife, and needle in the carbolized water, and make a longitudinal cut through the skin down to the sac, but do not injure it. Then push back the parts, and with the knife freshen the edges of the muscle; now after taking two stitches (more if the opening is huge) directly over the sac, draw the edges together, and cut the silk close; sprinkle in a little iodoform, draw the skin together with a stitch or two, and the work is done. If the puppies are as vigorous as those I have operated upon, it will probably be found next morning that the external stitches are out, and also the skin so swollen that the little fellow has been unable to reach the deep stitches, the all important point.

"Keep the puppy in a warm, clean place, but if any dirt should get in wash it out with warm water and castile soap, and sprinkle on a little more iodoform.

"But now about the deep stitches which have been allowed to remain in the muscle. Let them alone, for they cause no trouble as they become incysted. In two weeks the cure should be complete, and no scar, or at least, a very trifling one remains."

SPRAINS.

Sprains are of frequent occurrence among dogs, and often demand careful treatment. When severe they are attended with acute pain, heat, and swelling; with subsequent weakness and stiffness. When a large joint is affected there is often considerable constitutional disturbance, fever, rapid pulse, etc.

Perfect rest to the affected limb is the most essential measure, and if possible a splint should be used in its support. During the inflammatory stage cold water is the best application; the bandages being kept constantly wet. After the inflammation has subsided, the swelling may be reduced by slightly tightening the bandage, using uniform firmness and pressure.

If stiffness is not rapidly recovered from after exercise is allowed, stimulating liniments are to be used. It matters but little which is selected, as the efficacy of all depend almost entirely upon the hand-rubbing employed in their application; one part to three of the liniments ammonia and camphora make a good preparation.

BURNS AND SCALDS.

Extensive burns are always serious even if but superficial. The severe pain is exhausting; the internal organs are liable to become affected, and the healing stage is long and debilitating. It is important that some relieving application be made as soon as possible to a burned or scalded part, and for a short time at least domestic remedies must be relied upon. Among them a solution of common baking-soda is most effectual in relieving the acute pain; when the burn is only superficial or not severe it will remove it entirely in a very short time. It has the advantage of cleanliness, is always at hand, and if applied at once it will in a great measure prevent blistering, and the destructive changes in the skin. All that is necessary is to cover the injured parts with a thin cotton or linen cloth, and keep it constantly wetted with the soda lotion to prevent its drying. The relief felt from this application is immediate, and in many cases the acute pain subsides in less than half an hour. This solution may be persisted in for two or three days, and then the oxide of zinc ointment can be substituted.

If the injury is a severe one, and large blisters are formed, it will be well to use the soda until " carron oil " can be prepared; this is a liniment of equal parts of linseed oil and lime water; a very offensive preparation but highly effective. When obtained, the blisters should be pricked with a needle, and the whole part covered with a linen cloth soaked in the oil; then wrapped in soft cotton wool, to preserve the injured skin from the air, from cold, and to smother the bad odor. After the second day this covering may be removed, poultices of bread and water applied, and persisted in until suppuration is established; after which the oxide of zinc ointment may be employed until a cure is completed.

The strength of the injured animal must always be sustained by nutritious diet, and if need be stimulants.

If the burn is deep and much of the epithelial tissue and the hair-follicles destroyed, permanent disfigurement will result.

FRACTURES AND DISLOCATIONS.

The subject of fractures and dislocations is one which could with propriety be omitted from this work; for to those only who have made the anatomy of the dog a study, and who are perfectly familiar with his bony structure can an exhaustive treatise on the subject be comprehensive.

For other than a skilled surgeon to attempt the treatment of a fracture or the reduction of a dislocation, he must the more often grope about in the dark, flounder in the mire of uncertainty, inhumanly torture his helpless friend, and fail entirely in his purpose to repair the injury, or at best leave him permanently deformed. For this reason it is apparent that the discussion of the procedures of treatment is not only useless but pernicious, inviting attempts to do what only a surgeon should undertake.

To know when a fracture or dislocation does exist is important, therefore diagnosis and a brief consideration of the subject generally is admissable.

Fracture is said to be simple, when there is no wound of the skin communicating with the broken part; compound, when there is such a wound; comminuted, when the bone is broken in several fragments. In young animals fracture is sometimes partial, part only of the fibres breaking and the rest bending; to such the name green-stick or willow fracture is given. The long bones are most commonly broken; but any other may give way to direct violence.

The symptoms of fracture are deformity; such as displacement, bending, shortening, or twisting. Unnatural mobility; as shown by grasping the two ends of a broken bone and moving each independently of the other, or the yielding of a part on pressure. Crepitus; a grating heard or felt when the broken ends are rubbed against each other.

Of these three signs deformity is often absent in fractures of the ribs, pelvis, and shoulder blade; crepitus is prevented when the broken ends are displaced, and can only be felt when they are drawn into their natural positions. In addition to these symptoms there is more or less pain, swelling, and helplessness of the injured part. Dislocations are characterized by deformity and displacement. The external appearance of the joint is changed, the prominence disappears or moves to another part, usually leaving a depression in its place. A dislocated limb may be longer or shorter than the normal, according as the head of the bone is displaced upward or downward. It loses its mobility; can no longer be moved about freely and in as many directions. Movements cause pain, and some swelling occurs.

To distinguish between fracture and dislocation is rarely difficult; in the latter crepitus is absent, the bone can be moved less freely than natural, while in fracture mobility is increased. If a broken bone is drawn into its proper shape and position the deformity then disappears, but returns as soon as extension is discontinued; when a dislocated bone is drawn into place it will usually remain there.

When your dog has been injured, and the symptoms of fracture or dislocation are apparent or suspected, seek the aid of a surgeon as you would do had an accident of equal severity befallen any other near and dear friend.

INDEX.